The Construction of Horizontal and Vertical Water Wheels

by William Cullen

with an introduction by Roger Chambers

This work contains material that was originally published in 1871.

This publication is within the Public Domain.

*This edition is reprinted for educational purposes
and in accordance with all applicable Federal Laws.*

Introduction Copyright 2017 by Roger Chambers

Self Reliance Books

Get more historic titles on animal and stock breeding, gardening and old fashioned skills by visiting us at:

http://selfreliancebooks.blogspot.com/

Introduction

I am pleased to present yet another title on Homesteading and Farm Life.

The work is in the Public Domain and is re-printed here in accordance with Federal Laws.

As with all reprinted books of this age that are intended to perfectly reproduce the original edition, considerable pains and effort had to be undertaken to correct fading and sometimes outright damage to existing proofs of this title. At times, this task is quite monumental, requiring an almost total "rebuilding" of some pages from digital proofs of multiple copies. Despite this, imperfections still sometimes exist in the final proof and may detract from the visual appearance of the text.

I hope you enjoy reading this book as much as I enjoyed making it available to readers again.

Roger Chambers

To SIR ROBERT KANE,

PRESIDENT OF QUEEN'S COLLEGE, CORK, ETC.

Sir,
                Armagh, April, 1860.

As you were the first person to direct the attention of Irish mill-owners and Irish mechanics to Rülhman's work on the Turbine, and to express a hope that some practical man in this country might take up the subject, I venture to act on your suggestion, and also to respectfully dedicate the following pages to you as a slight tribute of my respect for your eminent talents, and my gratitude, in common with that of the working classes, for your zealous and untiring efforts to develop and diversify the industrial energies of the Irish people.

    I am, Sir,
      With great respect,
        Your obedient Servant,

          WILLIAM CULLEN.

PREFACE TO THE SECOND EDITION.

As the first edition of my 'Treatise on the Turbine' is exhausted, I have been induced, at the solicitation of numerous millwrights and engineers, to prepare a second edition for the press. I have therefore, in accordance with their wishes, devoted such leisure hours as I could spare from my professional avocation to a careful examination and revision of the first impression and its supplement, introducing such improvements as a closer observation and a more matured judgment suggested. For this end I have given new rules of construction, with diagrams and additional plates, which I flatter myself will impart new improvements and simplify its working. The great number of turbines that have been erected in the United Kingdom since I wrote the first treatise on this subject is a satisfactory evidence that they are gradually gaining ground, and that the milling public appreciate the merits of this ECONOMICAL MOTIVE POWER.

I therefore hope that the perusal of this second edition will be the cause of promoting a more extensive application of this important machine.

WILLIAM CULLEN.

Newry, August, 1871.

CONTENTS.

	PAGE
MOTIVES AND OBJECT OF THE AUTHOR	11
REMARKS ON TURBINES GENERALLY	17
THE TURBINE OF M. PONCELET DESCRIBED	17
THE TURBINE OF PROFESSOR THOMPSON, ESQ., DESCRIBED	17
THE TURBINE OF M. FONTAIN DESCRIBED	18
THE TURBINE OF M. JONVAL DESCRIBED	18
THE TURBINE OF ST. MAUR FLOUR-MILL	19
MESSIEURS CALLONS' IMPROVEMENT ON TURBINES	20
M. FOURNEYRON'S FIRST TURBINE	20
M. MORIN'S OPINION OF FOURNEYRON'S TURBINE	20
ELEMENTARY REMARKS	21
THE AUTHOR'S REMARKS	21
FOURNEYRON'S REMARKS ON HIS CURVES	23
ADVANTAGES OF THE TURBINES OVER VERTICAL WATER-WHEELS	24
THE TURBINE OF M. FOURNEYRON DESCRIBED. PLATE I.	26
FOURNEYRON'S TURBINE AT ST. BLAZIEN. PLATE II.	28
FOURNEYRON'S CURVES CONSTRUCTED. PLATES III. AND IV.	29
THE TURBINE GETTING THE WATER FROM THE UNDERSIDE. PLATE V.	30
A SECOND DESIGN FOR A TURBINE HAVING NO SLUICE. PLATE VI.	31
THE AUTHOR'S PROPORTIONS FOR TURBINES. PLATE VII.	32
EXAMPLE FOR GENERAL ILLUSTRATION	33
TO DELINEATE A TURBINE. PLATE VIII.	35
TO DELINEATE A TURBINE (2ND EXAMPLE). PLATE IX.	36
DESIGNS FOR SUPPORTING TURBINE SHAFTS. PLATE X.	37
TABLE OF SQUARE AND CUBE ROOTS TO ABRIDGE CALCULATION	39
TO ASCERTAIN THE VELOCITY OF WATER IN RIVERS	41
TABLE OF SURFACE AND MEAN VELOCITIES OF WATER	42
SECOND AND MORE CORRECT METHOD. PLATE XI.	42
THREE EXAMPLES OF SAME	43
TABLE OF THE DISCHARGE OF WATER OVER WEIRS	44
TABLE OF VELOCITY OF WATER THROUGH SLUICES	45
ON VERTICAL WATER-WHEELS	46

CONTENTS.

	PAGE
Table of Undershot Water-wheels	47
Common and Improved Breast-wheels	48, 49
Table of the Depth of Shrouding	50
Overshot Water-wheels	51
Example for their Velocity and Proportions	53
Formation of Water-wheel Buckets. Plate XI.	53
Description of Water-wheels. Plate XII.	54
Form of Wood Buckets	55
Table of the Diameter of Spur-wheels	56
Explanation of these Tables	62
Table of the Strength of Cast-iron Teeth	63

TREATISE.

MOTIVES AND OBJECTS OF THE AUTHOR.

I WISH, in the outset, to explain and justify my motives for appearing in a character which some few unreflecting readers may consider incompatible with my ordinary avocations.

In my professional employment as Millwright I frequently have had to construct vertical water-wheels, such as are commonly used in this country, and often thought that if plain rules were established instead of speculative theories, that more efficient machines would soon replace them. Endeavouring to supply this want, I now propose to offer to those interested in the construction of such machines my simple plans and calculations for such wheels.

After reading Sir Robert Kane's invaluable book on the 'Industrial Resources of Ireland,' I purchased his translation of Rülhman's work on Turbines, or horizontal water-wheels, a work which the translator represents as the most practical he could have selected. I applied myself to the study of this book, and found that, as regarded practice, it presented in its examples so many intricate and complicated formulas, as to deter many operatives, unless those of superior education, from even attempting to study it; or, in other words, for plain simple directions it is very deficient; mine, on the contrary, is intended to convey to such as are unable to comprehend those formulas, such plain and practical rules as will enable them to calculate and construct the various portions of the turbine water-wheels, which will realize the many advantages justly ascribed to these most important machines, the superiority of which wheels over the vertical now in use will be fully exhibited in the course of the following pages.

Trusting that I have now sufficiently explained my motives for appearing in print, and also the end which I have in view, I proceed to the narrative of my endeavours to qualify myself in some measure for the undertaking.

In order to arrive at the fountain-head, I visited France, and lost no time in securing an interview with the eminent inventor of the turbine, M. Fourneyron, in company with a gentleman then a resident in the Irish College at Paris, who kindly acted as interpreter; but from the extravagant demands made by M. Fourneyron for the information required, I found it impossible to effect any satisfactory arrangements with him. Being determined that my journey should not, if possible, prove abortive, I ascertained the address of his model maker, M. Clair, and visited his establishment, where I had an opportunity of inspecting a model turbine on M. Fontain's principle of construction, and was informed that this model was made to represent that of a large wheel, now driving a saw-mill at St. Maur. I subsequently visited the foundry of Messieurs Pihet and Co., where Fourneyron had his wheels manufactured, but information was there scrupulously withheld, and it was stated to us that the important portion of the constructive drawings could not be seen without the special leave of the inventor. While at the foundry, I had, however, an opportunity of examining the first wheel that was made for St. Blazien by Fourneyron, which was lying in the foundry yard. I then formed an idea of his plan of construction. I was also informed that, although this wheel, then lying in such a careless manner, had for a long time been in operation under a waterfall of about 118 feet, and had given the owner such satisfaction that he afterwards enlarged his mill and had this wheel replaced by a second turbine, made to suit a waterfall about three times higher than the former. I afterwards visited the flour-mill of St. Maur, and was kindly permitted to examine all its machinery, but the turbines at the time of my visit were working so deep in the back water, that I could only observe the upright shafts. A little higher up on the same river I examined the saw-mill before alluded to, and driven by Fontain's turbine, which was also working under back water, and at the same time driving a great number of upright and circular saws with undeviating regularity. The machinery in both these mills are driven by the river Marne. The commanding structure and appearance of the flour-mill, and its powerful and ingeniously constructed machinery, certainly excited my surprise and admiration. This mill has forty pairs of stones with all their requisite machinery for cleaning and dressing the flour, being driven by four turbines of co-equal power, each having on its shaft one large spur-wheel, gearing into ten pinions, which set in motion the ten pairs of stones surrounding the shaft, and the entire machinery of the mill, when in motion, resolves itself into harmonious uniformity of action. This plan, apparently so simple in construction and so grand in results, is worthy of the high fame of M. B. Fourneyron, who arranged and furnished the constructive drawings of the entire building and machinery.

My next visit was to the flour-mills of M. D'Arbley, at Corbeil, about sixteen miles from Paris, but admission was refused. I may here mention that these mills are four in number, originally driven by four vertical wheels, well proportioned, and about 18 feet in diameter, each mill containing ten pairs of millstones. I moreover learned that the proprietor of this extensive establishment, on hearing of the superior power of the turbines working in a similar waterfall at St. Maur, was induced to substitute a turbine in lieu of a vertical wheel, and after giving it a trial, which proved most satisfactory, he discontinued the use of the other three vertical wheels, which were immediately replaced with turbines similar to the one that he submitted to trial. Thus we see the proprietor of an extensive establishment substituting the turbines for the vertical wheels (with the performance of which he had been hitherto well satisfied).

On my return to Paris I was favoured with an interview with one of the eminent Company of Engineers, the Messrs. Callon and Sons, who with marked courtesy not only explained their plan of constructing turbines, but presented me with a printed report of several experiments made on the turbines erected by them at the flour-mill situated near Tillières.

I subsequently had the pleasure of being introduced to M. Passot, an eminent lecturer on hydraulics, and also on the construction of a new plan of turbine invented by himself. He kindly allowed me to inspect two of his model turbines, and expressed his great willingness to show all their parts in detail, which kindness he performed in that easy and polite style so peculiar to a Frenchman, although in many other cases I had to encounter that annoyance which is inseparable from the refusal of a favour: and I may here remark that I have found the inhabitants of that country, when a favour is sought from them, inclined to procrastinate, and require, particularly from a stranger, the favour of a "second call," that they may in the interim ascertain if possible the object of his inquiries or the purport of his mission: yet the suspicion of a stranger is not incompatible with the generous hospitalities of the French.

I trust that I have now, in the foregoing explanation, sufficiently laid before the reader the object of my visit, and that under the many disappointments I experienced I have made every conceivable effort to arrive at practical information in the construction of those invaluable machines.

On my return to Ireland, after due reflection and study, I committed to paper all the useful information I had collected in France; and in order to supply any deficiencies in so doing, I entered into correspondence with French engineers of high repute, who, in a great measure, very kindly supplied them. I then made a model turbine, according to Fourneyron's system, and the result of experiments I made on it was so satisfactory, that I deemed myself fully adequate for the task of constructing one on a large scale. I subsequently laid my plans before Mr. MacAdam, of Belfast, one of the proprietors of a foundry there, who, on considering the matter, requested me to furnish him with constructive drawings for the manufacture of a turbine; to which I assented, on certain conditions unnecessary to specify here. I furnished him with the requisite drawings, from which a wheel was constructed for a fall of 21 feet, erected at the linen bleach-mill of Messrs. Barklie and Co., at Mullamore, near Coleraine, in the county of Antrim.

This wheel had, on several occasions, worked nearly 9 feet under the tail water with the greatest uniformity, at the same time lifting a series of weighty stampers for washing linen cloth, and driving other machinery. Since that time the Messrs. MacAdam have made several wheels on Fourneyron's system. The work performed by one of them is fully reported in the 'Practical Mechanics' Journal' of April, 1853, in which it is stated that this turbine has produced a moving force equal to 85 per cent. of the absolute power of the water that supplied it.

Encouraged by so gratifying a result of my first effort, and with a further hope of acquiring more enlarged information, I revisited France; and being then more conversant with the usages and dispositions of the inhabitants, I was more fortunate in obtaining the object of my ambition than on the former occasion.

I was successful in obtaining access to the places to which I had been previously refused, and had the opportunity of inspecting all the parts of turbines in detail. In the Conservatoire des Arts, in Paris, I saw a model of one of the turbines that is at work in the flour-mill of St. Maur, as also models of proposed improvements on this wheel by other engineers. I inspected several other kinds of water-wheels—some were made to receive their impelling power at the exterior circumference, some others at the bottom, and to work at various angles. In this grand museum are exhibited machines remarkable for artistic taste and finished workmanship, and allowed by connoisseurs in this department to surpass any similar collection in Europe.

Being now fully satisfied that I had acquired all the information which I had long sought, I returned to Ireland with a resolve to convert that information to practical account.

I have thus detailed the means by which I obtained information for the construction of the turbine in its various forms, a subject on which I wrote in the year 1850, but not having then sufficient confidence in myself as to my capability of appearing as an author, I abandoned the idea of publishing my plans, hoping from time to time that some other individual more competent would undertake and publish a work on the subject. Finding that these hopes have not been realized, I now venture, humbly and diffidently, to present this small treatise to the public, but especially to artisans of my own class; and should it be found useful in tending towards the larger application of water-power, with which this country abounds, I shall consider myself well rewarded by the undertaking.

WILLIAM CULLEN.

ARMAGH.

REMARKS ON TURBINES GENERALLY.

I NOW deem it necessary to introduce a brief description of M. B. Fourneyron's turbine, and those of other engineers, and to advert to the manner in which the latter differ from the former.

The turbine of Fourneyron (as seen at Plate I.) is a horizontal water-wheel, consisting of a number of curved malleable-iron buckets, enclosed vertically between flat rings, which are connected with a disk-shaped flange of cast iron, secured to an upright shaft, by which the motion for driving any portion, or all the machinery, is transmitted. The impelling water is directed against the buckets of the wheel by means of channels formed all round the bottom of a cast-iron cylinder, which is firmly suspended within the centre of the revolving wheel; the water, by passing through these fixed channels, issues horizontally against the concave surface of the passing buckets, and after exerting its power in giving the wheel a rotary motion, escapes all round its exterior circumference into the wheel pit. When a turbine is applied to work under a high waterfall, the fixed cylinder is closely covered on the top, and the shaft passes through a stuffing box in its centre; and this cylinder has the impelling water let into it through a flanged pipe secured to its side; but with a low waterfall the cylinder is open at the top to admit the free flow of water from the millpond.

The plans of these wheels will be more fully explained hereafter by reference to the drawings or diagrams by which this work is illustrated.

M. Poncelet's turbines receive the impelling water through fixed channels, tangential to the outer circumference; and as the buckets recede from its pressure, the water gradually flows toward the centre of the wheel, and escapes through the channels at the inner circumference.

Professor Jas. Thompson, C.E., Belfast, has patented a turbine wheel, likewise with exterior injection, which he calls the vortex wheel, and of which a great number have been manufactured, and are working satisfactorily, both under high and low waterfalls. This patented wheel contains two tiers of radiating channels, placed in a fixed chamber, into the circumference of which the water is tangentially injected, and by means of curved guide-blades the quantity and direction of water are regulated and guided into the radiating passages of the wheel; and after gradually flowing through these channels, the water makes its exit at the inner circumference by two openings, one of which discharges upwards and the other downwards into the tail race.

M. Fontain's turbine has the water directed into it from the top side, by means of curved iron plates, suitably inclined and fixed in a stationary cylinder; the water passes through the channels formed by these plates, and directed against the reversed bent plates of the revolving wheel, which are situated under and exactly concentric with the fixed cylinder, the water enters the wheel with a velocity corresponding to the height of the water in the reservoir above it, and while passing from the top to the bottom of the wheel, expends all its moving force on the reversed bent plates of the moving wheel, and then escapes into the tail race.

The turbine of M. Jonval, like that of Fontain, receives the water on the top side, and issues underneath; but he has also discovered that he can place his wheels at any convenient distance less than 28 feet from the surface of the tail race, without losing any portion of the power of the waterfall; the suction occasioned by the pressure of air, removed from the underside of the wheel, he found to be exactly equal to the pressure of a column of water whose height was equal to the distance from the bottom of the wheel to the surface of the water in the tail course. This important discovery has also another useful feature in its construction, as, for instance, in case of repairs, this wheel can be made instantly dry and accessible; whereas submerged wheels, in order to be repaired, the surrounding water must be pumped out of the wheel pit.

Before M. Fourneyron's invention of the foot-step, he found himself greatly annoyed by having to pump out the water in case of disarrangement or wearing away of the pivot, which, according to report, was of frequent occurrence. He then constructed his pivots in such a manner that neither sand nor gravel could get near them, and this he effected by having a very hard steel thimble secured in the hollowed extremity of the shaft, which revolved on a fixed foot. This foot was so enclosed by a collar that surrounded it, that both the water and sand were completely excluded; but there is a cavity in this enclosure into which oil is admitted through a tube connected with a small pump, which forced the oil into radial grooves that are in the slightly convex extremity of the foot; and by this means the oil was kept constantly spread over the rubbing surface. This arrangement he found to work for nearly three years without repairs. In recent improvements oiling the foot-step under the water is rendered unnecessary, as experience has shown that water answers as well to all descriptions of pivots, and moreover that a pivot of lignum vitæ, carefully protected from sand or gravel, will, it is ascertained, outlast for years either brass or steel.

By various experiments on Fourneyron's wheel, it was proven that the greatest proportion of power obtained was when the spaces between the buckets were full of water, and that there was less proportionate power and a greater discharge of water as the sluice aperture became narrow. This deficiency of useful effect was evidently occasioned by the water after entering the wheel having room to spread out from its directed line of pressure. In order to remedy this defect, M. Fourneyron introduced horizontal plates into the wheel, which divided the depth of the buckets into different rows of compartments, making their distance from each other to correspond with the quantity of water that might be required to pass through them, so that if the supply was not sufficient to fill all the buckets, it might fill one or two tiers, leaving the others empty. Each of his wheels at St. Maur flour-mill was made to pass 80 cubic feet of water per second, which might be required at the time of floods; but I was informed that when the waterfall was at the height of 7 feet, 68 cubic feet per second were found to be quite sufficient to drive one of these turbines with ten pairs of stones, and all the cleaning and bolting machines belonging to it. The external diameter was 6 feet 7 inches, and made forty-five revolutions per minute, the vertical depth of the openings of the buckets 9½ inches; this depth was divided into three rows of buckets, thus: the top and lower ones were each 3½ inches, and the centre one 2½ inches, thirty-two buckets being in each row; yet it may be here mentioned, that although the introduction of the dividing plates in the buckets was found to be a decided improvement, where the supply of water is variable, still the inside edge of those plates prevented the free entrance of the water into the wheel, which defect was subsequently ingeniously remedied by the Messieurs Callons, of Paris. Those eminent engineers placed the sluice of their wheels inside of the dividing plates, and made horizontal divisions in them, which exactly corresponded in thickness, and on the same level as the divisions between the buckets; the entering water by these means met with no obstruction in passing to the wheel.

M. Fourneyron erected his first turbine at a place called "Pont-sur-l'Ognon," in France, and such were the difficulties and prejudices he had to contend with, that a period of seven years elapsed before he constructed a second in Franche-Comté, at the ironworks of M. Caron. This wheel was constructed for 8-horse power, to be driven by a waterfall of 4 feet 3 inches. When it had got a fair trial, its performance was so encouraging that the proprietor ordered another turbine of 50-horse power, to replace two of his vertical wheels, and in a short time afterwards Fourneyron received several other orders for turbines of a similar description.

M. Morin, one of the most celebrated hydraulic engineers in France, made several experiments on different horizontal water-wheels, and after having minutely tested their efficiency, came to the conclusion "that Fourneyron's turbine, above all other machines, under the smallest fall, utilizes the greatest quantity of water, and the grand results accomplished by them must eventually effect a great change in hydraulic engines as prime motors."

French scientific writers have given horizontal water-wheels a high repute, and frequently published tables of their experiments, whilst in poor neglected Ireland the only publication on this subject that I know of was a report of experiments made by that most profound scholar and astronomer, the Rev. Dr. Robinson, of Armagh Observatory, on a turbine constructed by the Messrs. Gardner, of this city. The result of these experiments, and the manner in which they were conducted, are mentioned in the Proceedings of the Royal Irish Academy, vol. iv.

Having thus adverted to the different forms of these wheels, I consider it right, in a brief manner, to allude to their impelling power and speed.

Now, as water is the element by which turbines act, and as the motive power of that machine consists in the utilization of the element in question, the relative velocity and general characteristic of action should be known to every millwright aspiring to the proficiency of the construction of hydraulic machines.

Water consists of exceedingly minute particles, or atoms of matter, independent of each other, yet acting together in mechanical combination. Solid bodies, such as stone, have a natural tendency to press only in one direction, namely, towards the earth's centre, in obedience to the law or principle of gravitation; but water is not restricted to this tendency to press downwards, but presses equally in all directions; the pressure on each particle of surface being proportionate to the weight of the superincumbent column of water.

The system generally adopted by theoretical writers for ascertaining the velocity of water under any given head of pressure, is to calculate it in the same manner as they would the velocity of a stone falling a given distance, from the surface of the pressing water to the aperture of issue. As this theory has never been recognized as being exactly correct, the practical man has therefore to supply the deficiency by making use of such materials as he has collected from published theory, combined with the result of his own experience.

The friction of solid bodies depends on the pressure exerted, and the nature of the surface pressed, but the friction of water depends on the relative velocity and extent of the rubbing surface. I have noticed these elementary principles, in hope that they may be not only useful in themselves, but also induce intelligent millwrights to consult standard works, in which the properties of water in a quiescent state, as also in motion, are explained with minuteness.

These remarks may be applied to the resistance that water encounters before it can enter a turbine; and as friction interferes most seriously with the utilization of water-power, the constructor of these wheels should ever bear in mind that, in the formation of turbines, he must arrange all its constituent parts so as to reduce the cause of friction as much as possible.

On reading the report of the engineers employed by the French Government to make accurate experiments on the turbine at Inval, near Paris, it appeared evident that the useful effect of the wheel increased with an augmented expenditure of water, or, in other words, the utilizing effects of the water always increased with increased height of sluice. This difference was caused by the area of the wide sluice opening always advancing in a much greater proportion than the friction, because the rubbing surface of the water, while passing under the sluice, was no greater for the wide than it was for the narrow opening; hence the increased velocity and utilizing effects of the water always increase with the increased height of the aperture. I moreover concluded that there must be an additional power in the action of the particles of water upon each other while imparting their force on the buckets; and that report also shows that the most effective speed of the circumference of the wheel corresponded with the theoretical velocity due to half the height of the waterfall; but according to experiments made on wheels of a similar construction, driven by higher waterfalls, it was ascertained that this proportional velocity did not produce the best effect. I may here observe, that the rule whereby I find the best working velocity for any height of fall exactly agrees with practice, but not with recognized theory, except that of the velocity of a body moving through water is as the cube root of its power, as ascertained by experience; and I now trust that

REMARKS ON TURBINES GENERALLY.

this hint may be the cause of some persons discovering the law that regulates the interaction of particles of water in motion.

With respect to the manner of directing the water on the wheel according to mathematical calculation, the greatest moving force of a turbine would be obtained by causing the entering water to press upon its buckets in a perpendicular direction to its radius, and to leave them in a tangential direction to its outer circumference; but that plan cannot be adopted, because there would not be sufficient room for the water to enter the buckets, and it would also be very much impeded by leaving them in a tangential direction to its outer circumference. This being the case, it was found necessary to construct the directing curves and buckets in such a manner that the water would both enter and leave the wheel at suitable angles: the angles that I have proven by experience to have given most satisfactory results will be noticed in another page.

As to the formation of the curves, many learned mathematicians have investigated this subject, but as yet their results, so far as they have been published, give very little useful information to the practical millwright. The inventor, M. Fourneyron, said " that it was the practical determination of the curves, *derived from experience alone*, which led him to the solution of the question." Yet he has taken good care to reserve to himself the plan of forming those curves; however, for the information of such as are interested in those machines, I have delineated by section the wheel that he constructed for the cotton-mill at St. Blazien (see Plate III.), and also the curves of the turbine in St. Maur flour-mill (Plate IV.); by reference to these diagrams it will be seen that the form of the curves exactly agrees with the simple intersection of lines or ordinates, which will be hereafter fully explained; and as I have proved this method of forming the buckets to correspond minutely in form with Fourneyron's original curves of these wheels, one being for a low and the other for a very high waterfall, I therefore conclude that in all probability he formed his curves in this manner.

ADVANTAGES OF THE TURBINES OVER VERTICAL WATER-WHEELS.

IF a ponderous vertical wheel be applied to a very high waterfall, its diameter will be so large, and its revolutions so very few, that it must be connected with a great deal of auxiliary machinery to impart that rapid motion which is generally required in mills. The consequence is, that through the friction occasioned by this additional machinery, considerable water-power is uselessly expended. On the contrary, the turbine being comparatively so small and its revolutions so numerous in a given time, its motive power can be at once transmitted, thereby enabling the mill-owner to dispense with the erection of additional shafting and wheels, and at the same time ensuring a considerable increase of power, with a machine not subject to get out of repair. Moreover, what operates as a disadvantage in the ordinary wheels, contributes to the more efficient working of the turbine; for the higher the waterfall, the smaller, and consequently the less expensive, the turbine adapted to it; and also it is applicable on falls of water so high that the ordinary wheel cannot be used. Another great property of the turbine is its constant and uniform motion, which arises from the diffusion of the impelling water over *the whole of the circumference at the same instant*. This perfect uniformity of motion is a peculiar feature of the turbine. Not only the application of a great quantity of water with a small wheel, but the horizontal direction in which it revolves affords a practical advantage for driving millstones, especially on low falls, without the wheel requiring intermediate gearing. The turbine is capable of working under the back water as long as the surface of the fluid in the reservoir remains the highest, during which time it will produce a moving force proportional to the difference between these two levels, without a perceptible diminution of the useful effect, thereby evidencing that it is exempt from the casualties to which the vertical wheel is so often subject.

ADVANTAGES OF THE TURBINES OVER VERTICAL WATER-WHEELS.

If a turbine be connected to a steam-engine during the summer months, while water is scarce, it can be made to transmit the highest obtainable power from the quantity of water by which it may be supplied, and it can be made so large as to drive all the works in winter—fuel then dear, and water is in abundance—saving the expense of fuel, economising that liquid that commonly runs to waste, and giving sufficient time for any repairs that might be required on the steam-engine, clearly showing the great advantages of the turbine over the vertical wheel, for it could not be made so large as to receive the extra water in winter, without lessening the effective power of the smaller quantities in summer. Whereas the turbine is capable of working under the tail water, and of discharging the largest supply of water for which it was made, and can work any less quantity without sustaining any diminution of its percentage power.

If an undershot wheel be applied to a fall of 3 or 4 feet, the useful effect produced will not exceed 30 or 34 per cent. of the expenditure. If a more favourable situation be selected, when, for instance, the waterfall would be 6 or 8 feet, and where the water is made to act as much as possible by its own weight, the useful effect might be from 50 to 60 per cent.; the small percentage in the former case may be accounted for by considering the oblique direction in which the force of the stream acts on the floats, and the loss sustained by the water which escaped between the breast ark and rim of the wheel. There is another loss of power in all vertical wheels, by keeping them at a convenient height from the tail course, which is necessary in order that the water may have free room to escape from the wheel. It is moreover necessary that the water should descend a determinate distance to have the required velocity on entering it, and only one half of this fall can have its full effect; whereas every inch of the entire waterfall may be made available when applied to drive a turbine, and which would yield under any fall a power of at least 75 per cent. of the water passing through it.

When it is considered that the great obstacle to the establishment of manufactories along the coast of Ireland is the expense of fuel, or rather the agency of steam as the prime motor, the advantage of the turbine is at once evidenced, for it can be driven by the fall supplied by the ebb and flow of the tide, as ably demonstrated by Sir Robert Kane's valuable writings on the 'Industrial Resources of Ireland.'

It is a fact of vast importance that turbines, on the improved principle of construction, supersede in America, France, and many other parts of the Continent every other description of water-wheels, and after long experience of their superior working power in impelling machinery, however ponderous and complex, have been stamped with the approbation of the most eminent mill-owners and manufacturers. And I am firmly convinced that their general introduction into the United Kingdom is now a mere question of time, and that foreign rivalry will necessitate a change in *our* hydraulic machines.

Having now, as I trust, invested the turbine with an interest for readers generally, I shall proceed to a closer, more detailed, and more demonstrative exposition of that water-wheel.

DESCRIPTION OF M. FOURNEYRON'S TURBINE AT ST. MAUR FLOUR-MILL.

PLATE I., Fig. 1, exhibits an external and internal vertical section of this wheel, on a half-inch scale, which deviates in external appearance very little from the original now in actual use, but the important parts fully agree. In this figure similar letters point out the same portions of the wheel, &c. $a\,a\,a\,a$ the shrouding of the wheel, between which are three rows of buckets; $b\,b$ the dish-formed plate, on the rim of which the buckets are secured; this plate is keyed on the upright shaft $c\,c$, on which is a large spur-wheel gearing into ten pinions, which give motion to the millstones; $e\,e$ is a pipe suspended from the top framing, serving the double purpose of keeping the shaft from the water, and of being a fixture to which the inside plate d is firmly keyed; this plate prevents the water from pressing on the foot of the upright shaft, and serves as a floor on which the directing curves ss are secured; ll are stays for keeping this plate concentric with the wheel that surrounds it; gg is a strong frame of wood, to which is bolted the outer cylinder hh; this cylinder is truly bored to receive the sluice ii, which is made water-tight by a ring of strong leather j; kk are wooden blocks, stationed all round the inside of the cylinder sluice, which are made to fit between the directing curves ss so as to move between them with the sluice; these blocks were intended to cause the water to flow on the wheel in small streams, as if injected from horizontal bent pipes; n, the pivot and foot-step are oiled by means of the tube m, which conveys the oil into grooves cut on the upper surface of pivot and oil-box, thereby keeping the foot constantly lubricated; o is the lever for raising the foot-step when worn; the sluice ii in this section is raised as high as the lowest row of buckets. The arrows show the direction in which the water flows from the river on all the buckets opposite the opening. The sluice is regulated as follows: qq are iron rods having their lower ends secured to three brackets p, cast on the inside of the sluice; screws are formed on the upper end of these rods, and pass through the eyes of three pinions t, having corresponding female screws inserted in them; these pinions gear into the centre wheel uu, which revolves freely round the short tube secured on the top of the framing vv; this wheel also gears into the pinion w. On the shaft of this wheel are connected worm-wheels, and by the action of a hand-wheel on them, motion is transmitted to the pinions and screw rods, thereby causing the sluice to have a vertical movement at the command of the person who attends it; when the sluice is lowered till it meets the margin of the plate d, it encloses the directing curves, and prevents the impelling water from entering the wheel, but when the sluice is raised, the water passes through the channels upon the passing buckets with a velocity proportional to the height of the water over it, thus giving motion to the turbine and all the gearing connected to its upright shaft. The reservoir for conducting the water to the wheel is composed of carefully executed masonry, in which are inserted the wooden beams gg, sheeted with plank, for the purpose of supporting and securing in its place the large outside cylinder h; this sheeting forms that part of the watercourse which surrounds the cylinder.

Fig. 2 represents the relative position of the directing curves and buckets of the wheel. Similar letters on both figures represent similar parts, and the direction in which the wheel revolves is shown by the arrows.

FOURNEYRON'S TURBINE AT ST. BLAZIEN.

PLATE II., Fig. 1. This wheel drives a spinning mill belonging to M. D'Eichtal, and is situated at the Black Forest of Baden. The waterfall is 354 feet; the external diameter of the wheel is 27 inches, and contains 36 buckets and 36 malleable-iron directors, each bucket being $1\frac{1}{8}$ inch deep, and the turbine is stated to be equal to 70 effective horse-power.

The vertical section of this machine on the right-hand view exhibits the external, and on the left hand the internal, appearance of the wheel and its appendages. $aaaa$, the shrouding with its enclosed buckets; bb, the dish-formed plate, on the flat rim of which the buckets are secured; this plate is keyed on the upright shaft cc, which by spur-wheels drives a second upright shaft, that transmits motion by bevelled wheels to all the machinery in the mill; n, the pivot and foot-step, which can be raised or lowered by the bevelled key o; oil is supplied to the rubbing surface of the pivot through the tube m; the pipe e is suspended from the top frame of cast iron, to which it is securely fixed; this pipe encloses the water from the upright shaft, and terminates below with the dish d, on which the curved directors are formed; hh is the large water-cistern, securely bolted by the flanges kk to the framing, which is made to receive it. The underside of this cistern is formed to fit the external circumference of the sluice, which is kept water-tight by the packing at *j*. Three brackets are placed at equal distances round the sluice; to these brackets are attached the sluice rods (two of them being shown by the letters *p p*), and by gearing similar to that of the turbine at St. Maur the rods and sluice *i i* are moved at will; recesses are cut in the bottom of the sluice to receive the directors, and when the sluice is raised up it serves as a fixed cover for them. The turbine at St. Maur has its water-cistern open at the top to admit the water from the river, but the turbine now being described is closely covered, and the impelling water is admitted through the tube *f*, 18 inches in diameter.

Fig. 2 is the plan of the wheel and directing curves. Similar letters on both figures refer to the same parts.

THE MANNER IN WHICH M. FOURNEYRON'S CURVES MAY BE CONSTRUCTED.

PLATE III. This plate exhibits a segment of the wheel at St. Blazien, delineated ¾-full size, in order to show the manner in which its curves may be formed.

Let A be a point on the inner circumference from which the curvature of a bucket is to commence, and A E the breadth of shrouding on radius line to the point of buckets. Divide the line A E into four equal parts, and measure along the circumference the distance E F equal to five of these parts, then the point F will be the other extremity of the bucket; make the angle B A C equal to 15 degrees from a tangent at the point A, and the angle D A E equal to 5 degrees from the radius line A E; prolong the line A C till it meets the circumference at G; draw the line G H indefinitely through the point F, and prolong the line A D, formed by the angle of 5 degrees, until it intersects the point at H; now the curve of the bucket is to be described by ordinates within the triangular space F A H; divide the line F H into any number of equal parts, and draw lines from A to each point marked on the line F H, likewise divide the line A H into the same number of equal parts, joining all the points of division with lines to G; the curve passing through the points where the corresponding exterior lines intersect will be the required shape of the bucket, which will serve as a model to trace all the other buckets of this wheel. The directors are circular arcs, and the centres for describing them are formed by an angle of 20 degrees from a point on the radius of wheel, which point being the middle of two directors, as delineated in diagram.

Plate IV. is a segment of the wheel at St. Maur flour-mill, drawn on a scale of $2\frac{1}{4}$ inches to a foot, having one of its buckets delineated in the same manner as that of the wheel just now described, from which it will be seen that the curves in both cases agree with the ordinates formed by the given lines. Similar letters on this segment refer to the same parts as on that of the turbine at St. Blazien.

Fig. 1, Plate V., is a turbine which differs from those commonly constructed by M. Fourneyron by its having the impelling water let into it from the underside, which is productive of the important advantage of conveniently oiling and examining the foot of the upright shaft, and the machine being placed under the floor of the mill does not interfere with its connected gearing; G is a malleable-iron shaft, having a bearing at e; n the foot-step; l, a key for adjusting the foot. The buckets are enclosed between $a\ a\ a\ a$, and are a portion of the casting $b\ b$, which, when keyed on the shaft c, form the revolving wheel. The part $b\ b$ is made to fit and play freely round the box of the foot-step; this short bearing surface is introduced to prevent any lateral tremor that might come on the wheel; $o\ o$ are openings for oiling and examining the foot of the shaft; $s\ s$, the fixed channels which are cast on the top of the cylinder $n\ n$, and are constructed like those shown on the diagram, Plate IX. The sluice $d\ d$ is represented fully opened to allow the entire water to be conducted to the wheel, the power of which may be diminished at pleasure by the action of the screw and handle Q, that are connected by the rod and lever to the sluice $d\ d$, as clearly shown on this diagram. The water is conducted from the reservoir by the pipe P, and is pressed forward to the wheel in the direction of the arrows.

Fig. 2, Plate V., is another arrangement for a turbine shaft, which can be as readily examined and oiled as any other bearing in the mill. By reference to this figure it will be seen that the wheel is suspended from the top of the upright shaft, instead of running it on a foot-step at the bottom; c is the shaft; G G, a bridge with brass bushing, in which this shaft revolves; P P, top bridge, bored to receive the box $b\ b$, fitted with bell-metal bushing; $e\ e$, on which the shaft and wheel is suspended by the projecting collars $d\ d$. When the bushing or collars wear, the shaft can be raised by the action of the screws $a\ a$, on the box $b\ b$. This plan of turbine shaft has been in successful operation for many years in Mr. Hague's corn-mill at Cavan, Ireland.

Plate VI. is a modification of this plan. By examining this diagram it will be seen that neither sluice, lever, nor its connecting irons are required. The quantity of impelling water issuing on the turbine is regulated by the action of the hand-wheel and screw at A. By them the suspended shaft S and turbine T T can be made to move either up or down past the fixed water-channels G G, which form the top portion of the pipe $d\ d$; the direction of the arrows points out the course of the water to the wheel. The valve V is for shutting up the supply water in the pipe P. But both it and the channels for directing the water on the wheel are shown fully open, so that full power of the water is conducted to the wheel.

I have found, by my experience, that the most effective speed or number of feet the inner circumference of a turbine would pass per second, was 4·4 times the square root of height of the waterfall (providing the height did not exceed 38 feet), and when the fall was higher than 38 feet the velocity, to give the inner circumference, corresponded with the cube root of the fall, multiplied by 8·1. These velocities produced the greatest amount of moving power, and was equal to two-thirds of the velocity of the water while entering the wheel. When the load was taken off the wheel, and it had free liberty to move with the greatest speed that the water could give it, the proportion of these speeds was found to be as 7 is to 3. Consequently with these proportions, and at a right angle to the radius, I constructed the parallelogram H P S G, Plate VII., Fig. 2, and took the direction of the diagonal G P for the centre of each jet of water entering the wheel. This direction will prevent as much as possible any turbulent action the water might have in meeting the buckets; and as the speed of the wheel is two-thirds of the speed of the water entering it, perpendicular to the radius at Fig. 1 a second parallelogram D A C B is constructed, and the length of the diagonal C A taken for the length of a bucket measured on a straight line. But this bucket must be curved to impede the direction given to the water, as shown at Fig. 3, and a curved channel given to the entering water, as shown at Fig. 2, by the dotted segment N P.

The line A G is prolonged till it crosses the internal periphery of the wheel at Q; the line P M made perpendicular to P G. Then P Q is divided into four equal parts, and three of these parts are set off on the line P M, and M is taken for a centre to describe the arc P N, which represents the middle of the channel that conducts the water to the wheel.

THE AUTHOR'S PROPORTIONS FOR TURBINES.

Q The quantity of water in cubic feet per second.

H The height of the waterfall in feet.

P The horse-power of the water at 75 per cent. $= \dfrac{Q H}{700}$.

d The inner diameter of the wheel $= \sqrt{\dfrac{Q}{\sqrt[3]{H}}} + \cdot 1$.

N The number of buckets $= d \times 3 + 28$.

THE AUTHOR'S PROPORTIONS FOR TURBINES.

B The breadth of shrouding $= \dfrac{d \times 55}{N}$.

s The shortest distance between two buckets $= \dfrac{B}{4\cdot 5}$.

D The exterior diameter to point of buckets $= B \times 2 + d$.

A The sectional area in inches between all the buckets $= \dfrac{Q \times 60}{\sqrt{H} \times 2\cdot 18}$.

h The height of buckets $= \dfrac{A}{NS}$.

b The breadth of rim for directors $= S \times 2\cdot 8$.

r The radius for centre of directing channels $= D \times 3\cdot 6$.

v The velocity of inner circumference for low falls $= \sqrt{H} \times 4\cdot 4$.

V The velocity of inner circumference for high falls $= \sqrt[3]{H} \times 8\cdot 1$.

R The revolutions of wheel per minute $= \dfrac{V \times 60}{D \times \tfrac{22}{7}}$.

U The diameter of turbine shaft in inches $= \sqrt[3]{\dfrac{P \times 240}{R}}$.

Note.—A $= \dfrac{Q \times 60}{\sqrt{H} \times 2\cdot 18}$ for high falls; but A $= \dfrac{Q \times 60}{2\cdot 08}$ for falls under 38 feet. Power was gained by extending the shroud about ¼th its breadth past the buckets when the water leaves them.

To find the Power of the Wheel at 75 per Cent.—The cubic feet of water passing through the wheel per minute, multiplied by the height of the waterfall, and divided by 700, will show by the quotient the power of the wheel.

Example for General Illustration.

Given 100 cubic feet of water per second on a waterfall of 9 feet, required the proportions for a turbine in accordance with the foregoing rules, to be constructed (to be driven by 50 cubic feet, and 25 occasionally), and at the time of working with these supplies to produce at least 75 per cent. of useful effect.

$\sqrt[2]{\dfrac{Q}{\sqrt[3]{H}}} + \cdot 1 = 7\cdot 03$, the interior diameter in feet.

$d \times 3 + 28 = 49$, nearest number of buckets.

$\dfrac{d \times 55}{N} = 7\cdot 89$ inches, breadth of shrouding to point of buckets.

34 THE AUTHOR'S PROPORTIONS FOR TURBINES.

$\dfrac{B}{4\cdot5} = 1\cdot753$ inch, shortest distance between two buckets.

$B \times 2 + d = 8\cdot345$ feet, exterior diameter.

$\dfrac{Q \times 60}{\sqrt{H} \times 2\cdot08} = 961\cdot53$ inches, sectional area of bucket opening.

$\dfrac{A}{N S} = 11\cdot175$ inches, collected height of buckets.

$S \times 2\cdot8 = 5\cdot806$, breadth of rim of directors in inches.

$d \times 3\cdot6 = 25\cdot308$ inches, radius for directors.

$\sqrt{H} \times 4\cdot4 = 13\cdot2$ feet, velocity of inner circumference.

$\dfrac{v \times 60}{d \times \frac{22}{7}} = 34\cdot54$ revolutions per minute.

$\dfrac{77\cdot14 \times 240}{34\cdot54} = 8\cdot12$ inches, diameter of shaft.

$\dfrac{11\cdot175}{2} = 5\cdot5877$ inches high, first tier of buckets to pass 50 feet.

$\dfrac{5\cdot5877}{2} = 2\cdot793$ inches high for second and third tiers, each to pass 25 feet.

ANOTHER EXAMPLE.

Required the quantity of cubic feet of water per minute, and all the other dimensions necessary to construct a turbine that will have 34-horse power on a waterfall of 99 feet 2 inches.

$\dfrac{34 \times 700}{99\cdot16} = 240$ cubic feet of water per minute, or 4 per second.

$\sqrt{\dfrac{Q}{\sqrt[3]{H}}} + \cdot1 = 1\cdot029$, the interior diameter in feet.

$d \times 3 + 28 = 31$, nearest number of buckets.

$\dfrac{d \times 55}{N} = 1\cdot826$ inch, breadth of shrouding.

$\dfrac{B}{4\cdot5} = \cdot406$ inch, shortest distance between two buckets.

$B \times 2 + d = 1\cdot333$, exterior diameter in feet to point of buckets.

$\dfrac{Q \times 60}{\sqrt{H} \times 2\cdot18} = 11\cdot06$ square inches sectional area of openings between buckets.

$S \times 2\cdot8 = \cdot9288$ inch, for rim of directors.

$\dfrac{A}{N S} = \cdot888$ inch, height of buckets.

$d \times 3\cdot 6 = 3\cdot 694$ inches, radius of directors.

$\sqrt[3]{H} \times 8\cdot 1 = 37\cdot 478$ feet, velocity of inner circumference.

$\dfrac{v \times 60}{d \times \frac{22}{7}} = 715\cdot 49$ revolutions per minute.

$\sqrt[3]{\dfrac{34 \times 240}{715\cdot 49}} = 2\frac{1}{4}$ inches, diameter of shaft.

TO DELINEATE A TURBINE ACCORDING TO THE FIRST EXAMPLE.

DESCRIBE the interior and exterior circumference of the shrouding, as noted in the particulars already found (see Plate VIII.), which is a horizontal section of a portion of the wheel and directors, drawn on a scale $2\frac{1}{4}$ inches to the foot. Inside diameter $7\cdot 03$ feet, and outside $8\cdot 345$ feet to the point of the buckets; the relative position and curvature of which, as well as of the directors, are delineated in this plate in the following manner:—

The radius line A E is divided into two equal parts. Measure three of these parts from A to F, then A and F will be the extremities of the buckets. From A with an angle of 8 degrees down from the radius lay off the directrix A H, and the directrix F H forming another angle of 8 degrees in from a tangent on the outer circle at F. The curve of the bucket is to be found within the triangular space F H A in the manner following:—Divide A H into any equal number of parts, and F H into the same number of parts; from A draw ordinates extending to these divisions on the line F H, and from F draw ordinates intersecting the points of equal division on the line A H; through the intersections of successive pairs of ordinates trace the curve of the buckets, as shown in the diagram.

The thickness is made in such a manner that the narrowest space between two buckets will be exactly $1\cdot 753$ inch; the convex side is so formed that the free action of the water entering on or pressing against the concave side will be impeded as little as possible. The full size of the wheel with all its buckets are delineated according to

36 TO DELINEATE A TURBINE ACCORDING TO THE FIRST EXAMPLE.

these observations. Cores are made the shape of the space between the buckets and depth corresponding to the distance between the shrouds of each tier, all of which are made as accurate and as smooth as possible, and the several tiers of buckets with their shroudings cast as in one piece.

The curvature of the centre of one of the conducting channels, which is the mean direction that the water should enter the wheel, is an arc, the centre of whose circle is on the line J K, which forms an angle of 23 degrees down from the line J M (through the centre of the wheel); and the mean direction J N is formed by the radius of 25·308 inches. The divisions that form these channels are so shaped that the narowest spaces next the wheel are exactly equal to the openings between the buckets, at the section where the water leaves them. Each tier of conducting channels are made the same depth as that ascertained for the buckets, and likewise cast in one piece.

Plate IX. represents a segment of a wheel, proportioned according to the last example (fall 99 feet 2 inches), and the concave sides of the buckets are traced by means of a slip of stiff paper with one straight edge, the inner and outer extremity of the bucket F A being ascertained according to the rule given, and delineated in Plate VIII.

From the commencement of the bucket the line A H is made perpendicular to A M, and the line F C equal to the distance F A, and to form an angle of 11 degrees from the line F O, which passes through the centre of the wheel. This being done, the distance 1 and 2 is marked on the slip of paper equal to the distance A H, and the distance 1·3 equal to F H. Then the straight edge of the paper is placed on the line F C, and is made to revolve round to A, always keeping the point 2 on the line F C, and point 3 on the line A H; during this revolution, point 1 is on the required curve F A. This shape of bucket is supposed to gradually impede every effort that the water would make to flow in the direction in which it was pressed, and transmits a corresponding rotary motion to the wheel. The same directions that were given for forming the thickness of the convex side of the bucket on the large wheel should be strictly attended to in this, as well as all other sizes of wheels, likewise the thickness and shape of the plates that form the conducting channels, which should be nearly parallel, having the narrowest part of the opening next the wheel, and the points of the curves shaped so as to impede the free entrance of the water as little as possible, and all scales and roughness removed, having the centre of each stream of water running into the wheel, to form an angle of 23 degrees to the tangent of the inside circumference. When the height of the fall and quantity of water to be discharged are variable, the diameter of the wheel is calculated for the largest supply of water, and the distance between the shroudings proportioned so that the fluctuating supply would give the most useful effect throughout the greater part of the year, and the average height of the fall is calculated for the speed of the wheel. The conduit through which the water is brought to the wheel should be so large that the water would flow into the cistern at a slow velocity, and capable of supplying more water than the wheel could discharge. The upper shrouding to always be under the surface of the tail water.

DESIGNS FOR SUPPORTING TURBINE SHAFTS.

PLATE X., Fig. 1. S is a section of the lower portion of the turbine shaft for 40-horse power, on a waterfall of 8½ feet; P the lignum-vitæ pivot, on which the wheel revolves; D D the pivot-stand bored to receive it. This pivot is prevented from turning in the stand by the mortise cut across it to receive the lever L; this lever is for the purpose of regulating the height of the wheel, and is moved by the action of a screw and connecting rod, like the lever shown in the turbine of the flour-mill at St. Maur. In the hollowed end of the shaft T is secured the brass thimble which revolves on the stationary pivot, and is lubricated through the channel C, which has several other channels leading into it, so that an ample supply of water is spread over the entire rubbing surface of the wood and brass. This plan has been in successful operation for many years. It has been proved that pivots made of lignum vitæ are better and more durable for working under water than either iron or steel pivots.

Fig. 2 represents the construction of the foot and foot-step of a turbine shaft for 90-horse power, on a waterfall of 40 feet; S the shaft; D D the foot-stand, bored to receive the brass thimble $i\,i$. O is an aperture by which oil is supplied to the cistern C C; the heated air makes its escape through this aperture; B B are discs of brass, M a disc of cast metal, which are truly turned up, having a small hole bored in the centre of each, and radiating holes bored half into the brasses and half into the cast metal, and both discs made a good fit into the thimble; grooves are cut in the bored part of the foot-stand, and holes bored through the brass thimble, so that the oil may pass freely from the cistern C C through the several channels to lubricate the foot and discs. This plan has been found to work with very little friction, and it has been ascertained that if any extra friction takes place between the bottom of the foot and the top surface of the disc next it, some of the discs will revolve, but the continuous system of lubrication will in a few minutes cause the foot to glide over its former resistance and again revolve on the surface of the disc next to it.

By reference to Fig. 3, it will be seen that this shaft does not run upon a step at the bottom, but upon discs having openings in the centre of each to let the shaft pass through them. This diagram is for a turbine shaft of 60-horse power, on a fall of 7 feet; S the shaft; K a block of cast metal secured on it, and by which the shaft is suspended; N a screw and nut for regulating the height of the shaft; B B brass discs; N cast-metal disc, on which the cast-metal block revolves. These discs are truly turned up, and placed in the stand D D; radiating holes are bored half into the brasses and half into the cast metal to receive the oil which can flow freely through the radiating channels from the cistern C C, and lubricate the revolving surface; H is the cast-iron cover enclosing the oil chamber and revolving head of the shaft. This plan of bearing up turbine shafts has also given great satisfaction. The remark made on the discs in latter case may in like manner be applied to this construction.

SQUARE AND CUBE ROOTS.

A Table of Square and Cube Roots, from 1 to 200.

Num.	Sq. root	Cub. root	Num.	Sq. root	Cub. root	Num.	Sq. root	Cub. root	Num.	Sq. root	Cub. root	Num.	Sq. root	Cub. root
1	1·000	1·000	51	7·141	3·708	101	10·050	4·657	151	12·288	5·325			
2	1·414	1·260	52	7·211	3·732	102	10·099	4·672	152	12·329	5·337			
3	1·732	1·442	53	7·280	3·756	103	10·149	4·687	153	12·369	5·348			
4	2·000	1·587	54	7·348	3·780	104	10·198	4·702	154	12·409	5·360			
5	2·236	1·710	55	7·416	3·803	105	10·247	4·717	155	12·450	5·371			
6	2·449	1·817	56	7·483	3·826	106	10·295	4·732	156	12·490	5·383			
7	2·645	1·913	57	7·549	3·848	107	10·344	4·747	157	12·530	5·394			
8	2·828	2·000	58	7·615	3·871	108	10·392	4·762	158	12·570	5·406			
9	3·000	2·080	59	7·681	3·893	109	10·440	4·777	159	12·609	5·417			
10	3·162	2·154	60	7·745	3·915	110	10·488	4·791	160	12·649	5·429			
11	3·316	2·223	61	7·810	3·936	111	10·535	4·806	161	12·688	5·440			
12	3·464	2·289	62	7·874	3·958	112	10·583	4·820	162	12·728	5·451			
13	3·605	2·351	63	7·937	3·979	113	10·630	4·834	163	12·767	5·462			
14	3·741	2·410	64	8·000	4·000	114	10·677	4·849	164	12·806	5·474			
15	3·873	2·466	65	8·062	4·021	115	10·724	4·863	165	12·845	5·485			
16	4·000	2·520	66	8·124	4·041	116	10·770	4·877	166	12·884	5·496			
17	4·123	2·571	67	8·185	4·061	117	10·816	4·891	167	12·923	5·507			
18	4·242	2·620	68	8·246	4·081	118	10·863	4·905	168	12·961	5·518			
19	4·359	2·668	69	8·306	4·101	119	10·909	4·918	169	13·000	5·528			
20	4·472	2·714	70	8·366	4·121	120	10·954	4·932	170	13·038	5·539			
21	4·582	2·759	71	8·426	4·141	121	11·000	4·946	171	13·076	5·550			
22	4·690	2·802	72	8·485	4·160	122	11·045	4·959	172	13·115	5·561			
23	4·796	2·844	73	8·544	4·179	123	11·090	4·973	173	13·153	5·573			
24	4·899	2·884	74	8·602	4·198	124	11·135	4·986	174	13·191	5·583			
25	5·000	2·924	75	8·660	4·217	125	11·180	5·000	175	13·229	5·593			
26	5·099	2·962	76	8·718	4·236	126	11·225	5·013	176	13·266	5·604			
27	5·196	3·000	77	8·775	4·254	127	11·269	5·026	177	13·304	5·614			
28	5·291	3·036	78	8·831	4·272	128	11·313	5·039	178	13·341	5·625			
29	5·385	3·072	79	8·888	4·291	129	11·358	5·052	179	13·379	5·636			
30	5·477	3·107	80	8·944	4·309	130	11·402	5·065	180	13·416	5·646			
31	5·567	3·141	81	9·000	4·326	131	11·445	5·078	181	13·453	5·656			
32	5·657	3·175	82	9·055	4·344	132	11·489	5·091	182	13·490	5·667			
33	5·744	3·207	83	9·110	4·362	133	11·532	5·104	183	13·527	5·677			
34	5·831	3·239	84	9·165	4·379	134	11·576	5·117	184	13·564	5·688			
35	5·916	3·271	85	9·219	4·397	135	11·619	5·130	185	13·601	5·698			
36	6·000	3·302	86	9·273	4·414	136	11·662	5·142	186	13·638	5·708			
37	6·082	3·332	87	9·327	4·431	137	11·704	5·155	187	13·675	5·718			
38	6·164	3·362	88	9·381	4·447	138	11·747	5·167	188	13·711	5·728			
39	6·245	3·391	89	9·434	4·464	139	11·790	5·180	189	13·747	5·739			
40	6·324	3·420	90	9·487	4·481	140	11·832	5·192	190	13·784	5·749			
41	6·403	3·448	91	9·539	4·498	141	11·874	5·204	191	13·820	5·759			
42	6·480	3·476	92	9·591	4·514	142	11·916	5·217	192	13·856	5·769			
43	6·557	3·503	93	9·643	4·530	143	11·958	5·229	193	13·892	5·779			
44	6·633	3·530	94	9·695	4·547	144	12·000	5·241	194	13·928	5·789			
45	6·708	3·557	95	9·746	4·563	145	12·041	5·253	195	13·964	5·799			
46	6·782	3·583	96	9·798	4·579	146	12·083	5·265	196	14·000	5·809			
47	6·855	3·609	97	9·849	4·594	147	12·124	5·277	197	14·035	5·818			
48	6·928	3·634	98	9·899	4·610	148	12·165	5·289	198	14·071	5·828			
49	7·000	3·659	99	9·950	4·626	149	12·206	5·301	199	14·107	5·838			
50	7·071	3·684	100	10·000	4·641	150	12·247	5·313	200	14·142	5·848			

To find the square or cube root of a number containing decimals.

SQUARE AND CUBE ROOTS.

Subtract the square root or cube root of the integer of the given number from the root of the next higher number, and multiply the difference by the decimal part. The product added to the root of the integer of the given number will be the answer required.

EXAMPLE. Required the square root of $15 \cdot 52$.

$\sqrt{15} = 3 \cdot 873$ and $\sqrt{16} = 4 \cdot 000$, the difference $\cdot 127 \times 52 + 3 \cdot 873 = 3 \cdot 937$, the answer required.

SUPPLEMENT.

TO ASCERTAIN THE VELOCITY OF WATER IN RIVERS.

SELECT a portion of the watercourse where the direction of the running water is as straight as possible. Measure some convenient distance along the stream, say 100 feet; at the extremities of this length, and at right angles across the stream, fix two straight cords; then get a few floats of hard wood, or well-corked bottles (they should be so weighty that when placed in the water they would not project over it as to be materially affected by the wind). After having these preparations made the floats are dropped lightly into the current at a little distance above the upper cord, then note the time by a stop-watch that the float takes to pass over the distance between the two cords. This experiment should be repeated several times with floats both in the middle and near the sides of the stream, the arithmetic mean is then taken for the surface velocity of each experiment. Having by these means found the several spaces run over in a given time, the mean proportion of all these trials is taken for the surface velocity of the water. Four-fifths of the surface velocity is a good approximation to take for the mean velocity of the stream, or the velocity it would have supposing all the particles of the stream to have moved in every part of its channel with one uniform motion. If this velocity be multiplied by the breadth and depth of the mass of flowing water, the product will be the quantity of cubic feet which passed from one cord to the other during the time of observation. This may be adapted to any other proportion of time.

EXAMPLE. Required the quantity of water discharged in cubic feet per minute, in a channel 8 feet broad, and water 2 feet deep, supposing the measured distance between two cords placed across the channel to be 100 feet; and after making five different experiments on the time which floats took to pass from one cord to the other, it was found to be as follows:—At the first trial it took $11\frac{1}{2}$ seconds; at the second, 12; at the third, 13; at the fourth, 10; and at the fifth, 13½. Then $11½ + 12 + 13 + 10 + 13½ = 60$; and $\frac{60}{5} = 12$ seconds for the mean time that the floats took to pass the 100 feet, $\frac{4''}{5}$ this distance = 80 feet for the mean velocity of the stream in 12 seconds; 12 : 60 :: 80 : 400 feet for the velocity of the stream per minute; $400 \times 8 \times 2 = 6400$ cubic feet, the required answer.

In order to abridge calculation, I have annexed a Table by which the mean velocities corresponding to surface velocities from 120 to 800 feet per minute may be seen at one glance.

TABLE of SURFACE and MEAN VELOCITIES of WATER in FEET per MINUTE.

Surface Velocity.	Mean Velocity.	Surface Velocity.	Mean Velocity.	Surface Velocity.	Mean Velocity.	Surface Velocity.	Mean Velocity.	Surface Velocity.	Mean Velocity.
120	98·00	175	147·95	230	198·60	285	249·75	380	338·9
122·5	100·25	177·5	150·20	232·5	200·90	287·5	252·10	385	343·6
125	102·50	180	152·50	235	203·25	290	254·45	390	348·3
127·5	104·75	182·5	154·80	237·5	205·55	292·5	256·75	395	353·0
130	107·00	185	157·10	240	207·85	295	259·10	400	357·8
132·5	109·25	187·5	159·40	242·5	210·20	297·5	261·45	405	362·5
135	111·55	190	161·70	245	212·50	300	263·75	410	367·2
137·5	113·80	192·5	164·00	247·5	214·85	305	268·4	415	371·9
140	116·05	195	166·30	250	217·15	310	273·1	420	376·7
142·5	118·30	197·5	168·60	252·5	219·50	315	277·8	425	381·4
145	120·60	200	170·90	255	221·80	320	282·5	430	386·1
147·5	122·85	202·5	173·20	257·5	224·15	325	287·2	435	390·8
150	125·15	205	175·50	260	226·45	330	291·9	440	395·6
152·5	127·40	207·5	177·80	262·5	228·80	335	296·6	445	400·3
155	129·65	210	180·10	265	231·10	340	301·2	450	405·1
157·5	131·95	212·5	182·40	267·5	233·45	345	305·9	500	452·5
160	134·20	215	184·75	270	235·75	350	310·6	550	500·0
162·5	136·50	217·5	187·05	272·5	238·10	355	315·3	600	547·7
165	138·80	220	189·35	275	240·45	360	320·1	650	595·5
167·5	141·05	222·5	191·65	277·5	242·75	365	324·8	700	643·3
170	143·35	225	193·95	280	245·10	370	329·5	750	691·2
172·5	145·65	227·5	196·30	282·5	247·45	375	334·2	800	739·2

SECOND AND MORE CORRECT METHOD.

Look for some convenient portion of the watercourse, where the water is running slowly. In a perpendicular position and at right angles across it, fix an overfall of boards, as is shown by the diagrams

1 and 2, Plate XI., in such a manner that the water shall accumulate behind it so high that it will flow over the horizontal edge of the rectangular opening R P, and afterwards have ample room for its free escape. The edges of the boards forming the opening should be sharp and smooth, and all leaks round the bottom and sides carefully closed with clay, in order that the whole water may pass through the opening ef. The cross section of the water should be at least eight times the length of notch ef, and the depth of the water under C D should be four times the depth of B D. At a distance below the overfall a rod O is fixed vertically, having a mark D at the exact level of the edge of the notch C D. When the water has risen to its greatest height, and its surface observed to be horizontal, the depth from the surface of the water A B to the mark on the rod D is noted in inches in column the first of the following Table. Then on the same line with that number in column three of the Table is a number which if multiplied by ef, the width of the notch in inches, will give the quantity of water in cubic feet per minute, providing the width of the notch is less than seven times the depth B D, but when the width ef is more than seven times the depth that D is below B the second column is taken; when ef is equal to B D the quantity of water discharged through the notch will be as much less as the quantity contained in the third column of the Table is to that contained in the second.

1st EXAMPLE. When ef and B D are each 24 inches, the third column in the Table gives a discharge of 47·515 cubic feet of water in one minute, and the second column gives 50·465. The difference of these two numbers is 2·95. This difference subtracted from 47·515, the remainder will be 44·55 cubic feet of water discharged per minute by every inch in width of the notch when B D and ef are of the dimensions now given.

2nd EXAMPLE. Required the quantity of water in cubic feet per minute discharged through a notch 60 inches long and 9 inches deep from B to D. In that case, as the distance from B to D is less than seven times the distance from e to f, we take the third column to find the quantity. Then opposite 9, in column 1, the corresponding number is 10·91, and this multiplied by 60, the length, gives 654·6 cubic feet per minute.

QUANTITY of WATER DISCHARGED over WEIRS or NOTCHES in CUBIC FEET per MINUTE by every INCH of their LENGTH. Column 2 is calculated from SMEATON'S Experiments. Column 3 according to DU BUAT'S.

1. Depth falling over.	2. Discharge per Minute over Weir.	3. Discharge per Minute over Notch.	1. Depth falling over.	2. Discharge per Minute over Weir.	3. Discharge per Minute over Notch.	1. Depth falling over.	2. Discharge per Minute over Weir.	3. Discharge per Minute over Notch.
inches.	cubic feet.	cubic feet.	inches.	cubic feet.	cubic feet.	inches.	cubic feet.	cubic feet.
·25	·053	·50	7·25	8·379	7·889	16·5	28·768	27·085
·5	·115	·143	7·5	8·816	8·300	17	30·085	28·326
·75	·278	·262	7·75	9·260	8·719	17·5	31·422	29·585
·1	·429	·404	8	9·712	9·144	18	32·778	30·862
1·25	·599	·565	8·25	10·171	9·576	18·5	34·153	32·156
1·5	·788	·742	8·5	10·636	10·015	19	35·547	33·469
1·75	·993	·936	8·75	11·101	10·460	19·5	36·960	34·799
2	1·214	1·143	9	11·589	10·911	20	38·390	36·146
2·25	1·448	1·364	9·25	12·075	11·369	20·5	39·839	37·509
2·5	1·697	1·597	9·5	12·568	11·833	21	41·305	38·890
2·75	1·957	1·843	9·75	13·067	12·303	21·5	42·789	40·287
3	2·230	2·910	10	13·573	12·779	22	44·291	41·701
3·25	2·515	2·368	10·25	14·085	13·262	22·5	45·809	43·131
3·5	2·810	2·646	10·5	14·604	13·750	23	47·344	44·576
3·75	3·117	2·935	10·75	15·128	14·244	23·5	48·897	46·038
4	·3·434	3·233	11	15·659	14·743	24	50·465	47·515
4·25	3·761	3·541	11·25	16·196	15·249	25	53·652	50·517
4·5	4·097	3·858	11·5	16·739	15·760	26	56·903	53·576
4·75	4·443	4·184	11·75	17·296	16·277	27	60·218	56·697
5	4·799	4·518	12	17·926	16·799	28	63·594	59·875
5·25	5·163	4·861	12·5	18·969	17·860	29	67·050	63·111
5·5	5·536	5·213	13	20·119	18·942	30	70·528	66·404
5·75	5·918	5·572	13·5	21·299	20·045	31	74·083	69·752
6	6·308	5·939	14	22·484	21·169	32	77·698	73·154
6·25	6·706	6·314	14·5	23·699	22·313	33	81·367	76·602
6·5	7·113	6·697	15	24·935	23·477	34	85·094	80·118
6·75	7·527	7·087	15·5	26·192	24·661	35	88·875	83·678
7	7·949	7·484	16	27·470	25·864	36	92·711	87·290

When using the above Table the breadth of notch must not exceed six times the depth of the water running over it. The following Table is used for common weirs with straight approach, but when the approach is rough, a medium between these quantities is taken.

3rd EXAMPLE. Let BD the depth be 12 inches, and *ef* the length 100 inches, to find the number of cubic feet that would pass through this opening per minute. In this case the second column is taken. Opposite 12 inches deep, in the first, is 17·926, which multiplied by the length, 100 inches, will give 1700·926 cubic feet per minute. The second column is made in accordance with Mr. Smeaton's experiments, which were made on water running in parallel channels, and flowing over the edge of a plank placed horizontally and at right angles across the entire breadth of a weir or stream, offering no lateral obstruction to the water flowing over it. But the third column was calculated for a waste board, having a notch cut in it narrower than the channel in which it was placed, in consequence of which the water, in meeting the ends of the notch, and afterwards turning into the opening, contracts its passage, and causes a less discharge in a given time: the breadth of the notch and depth of the water being the same in both cases.

VELOCITY of WATER THROUGH ORDINARY SLUICES UNDER HEADS from 4 inches to 6 feet.

Head of Water from the Surface to Centre of Sluice Opening.	Velocity of Water when the Head of Pressure is less than 6 times the Depth of the Sluice Opening.	Velocity of Water when the Head of Pressure is more than 6 times the Depth of Sluice Opening.	Head of Water from the Surface to Centre of Sluice Opening.	Velocity of Water when the Head of Pressure is less than 6 times the Depth of the Sluice Opening.	Velocity of Water when the Head of Pressure is more than 6 times the Depth of Sluice Opening.	Head of Water from the Surface to Centre of Sluice Opening.	Velocity of Water when the Head of Pressure is less than 6 times the Depth of the Sluice Opening.	Velocity of Water when the Head of Pressure is more than 6 times the Depth of Sluice Opening.
ft. in.			ft. in.			ft. in.		
0 4	3·09	2·956	1 2	5·77	5·530	2 10½	9·06	8·681
0 4¼	3·18	3·047	1 2½	5·88	5·618	3 0	9·25	8·868
0 4½	3·27	3·135	1 3	5·98	5·716	3 1½	9·45	9·051
0 5	3·45	3·305	1 3½	6·07	5·719	3 3	9·55	9·133
0 5½	3·62	3·466	1 4	6·17	5·811	3 4½	9·82	9·305
0 6	3·78	3·621	1 4½	6·27	6·004	3 6	9·99	9·579
0 6½	3·93	3·868	1 5	6·36	6·094	3 7½	10·18	9·749
0 7	4·08	3·910	1 5½	6·45	6·183	3 9	10·35	9·915
0 7½	4·23	4·047	1 6	6·54	6·271	3 10½	10·52	10·079
0 8	4·36	4·181	1 7	6·72	6·443	4 0	10·70	10·209
0 8½	4·50	4·309	1 8	6·90	6·610	4 2	10·91	10·451
0 9	4·63	4·436	1 9	7·07	6·773	4 4	11·13	10·658
0 9½	4·75	4·555	1 10	7·24	6·933	4 6	11·34	10·861
0 10	4·88	4·674	1 11	7·40	7·088	4 8	11·55	11·060
0 10½	5·00	4·789	2 0	7·56	7·241	4 10	11·75	11·256
0 11	5·12	4·902	2 1½	7·80	7·464	5 0	11·95	11·449
0 11½	5·23	5·012	2 3	8·02	7·680	5 3	12·25	11·737
1 0	5·34	5·120	2 4½	·8·24	7·891	5 6	12·53	12·007
1 0½	5·45	5·223	2 6	8·45	8·095	5 9	12·82	12·278
1 1	5·56	5·327	2 7½	8·66	8·295	6 0	13·09	12·488
1 1½	5·67	5·431	2 9	8·96	8·491			

If the quantity of water that passes through a sluice opening be required, see the velocity of the water in the above Table corresponding to the given head of water, which, if multiplied by the area of the sluice opening, will be the quantity of cubic feet passing per second.

VERTICAL WATER-WHEELS.

VERTICAL wheels are commonly termed undershot wheels, breast-wheels, and overshot wheels, deriving their name from the position in which the propelling water enters them.

BRIEF OBSERVATIONS ON THESE MACHINES.

The undershot wheel is driven by the force of a stream of water acting on its buckets at the height of from 7 to 26 inches from the point where the water leaves the wheel. And the product of the square root of the perpendicular height in feet from the surface of the water in the head race till it meets the buckets, multiplied by 4·5, will be the velocity of the skirt of the wheel in feet per second, which speed I have found to give the best effect. Wheels of this class, with straight wood buckets, are cheap in construction, and are adopted when there is a large supply of water on low falls; but their effective power seldom exceeds 34 per cent. of the water that turns them. If iron curved buckets are substituted, the power is about one-fourth more. It would be necessary to set the sluice at the part where the water leaves it as near the wheel as possible, with a backward inclination from the wheel and its under edge so formed as to direct the water on the wheel at an angle of about 30 degrees from a tangent to the hem. The sides of the cistern which conducts the water to the wheel should be a little contracted next the sluice-frame, and the sluice opening about 4 inches narrower than the wheel. The framing of the sluice-gates ought to be placed in such a manner as to obstruct the running water as little as possible before it meets the wheel, for if it projects inwards past the sides of the cistern, the currents of water running towards the sluice opening, where they meet with this projection, must necessarily rush across the central column, diminishing its velocity and power. Therefore great care should be taken in the formation of the sluice and its frame, and placing them in their proper position, so that the impelling water will have a smooth channel for its passage before meeting the wheel; and the wheel should be made no wider than what will contain the water that is necessary to drive the greatest load that may be on it. If it be made wider, the sluice opening becomes shallower, thereby diminishing the velocity of the stream. In the following Table is calculated the proportions for undershot wheels with iron buckets, working under different falls, to be equal to 10-horse power:—

UNDERSHOT WHEELS.

Height of Fall from Surface of Top to Surface of Tail Water.	Head of Contact.	Velocity of Circumference in feet per second.	Diameter of the Wheel in feet.	Turns per Minute.	Number of Buckets.	Depth of Curved Buckets; if Straight, ⅓ broader.	Length of Buckets for 10-horse power.
2·7	1·764	5·94	14	8·04	30	1·33	6
4·0	2·642	7·27	15	9·26	32	1·63	4·93
5·3	3·528	8·40	16	10·02	32	1·88	4·27
6·6	4·506	9·49	17	10·65	32	2·10	3·7
7·9	5·294	10·29	19	10·34	36	2·30	3·5
9·2	6·165	11·10	—	10·60	40	2·49	3·22

If a wheel is required for any other power on a given waterfall, it is to be proportioned according to the above Table; for example:—

Required the speed and proportions of a wheel to be equal to 10-horse power on a waterfall of 7·9 feet. See 7·9 feet in first column of Table. Then on the same line is 19 feet diameter, 10·34 revolutions per minute, 36 buckets, 2·3 feet deep, and 3·5 feet long.

But if a wheel is required for 20-horse power, instead of 10, the length of the buckets is to be multiplied by 2, and so on for different proportions of power.

BREAST-WHEEL.

The ordinary breast-wheel, like the undershot wheel, has its power transmitted from the axle. It partakes of the properties both of an undershot and overshot wheel, and is applied on waterfalls from 7 to 14 feet high, the water acting on it partly by weight and partly by impulse, and is discharged from under the sluice. That part of the fall before the water enters the buckets is by impulse, and the remaining portion by weight. Wheels of this class have a velocity at the circumference of 6 to 8 feet per second, and should have a diameter of at least 16 or 18 feet. The large diameter moves through the back water easier than the small, and the water in it has a less angular velocity.

Particular attention should be paid in allowing sufficient room for the air to escape from the buckets, which may be done by making the stream running into them narrower than the wheel, or leaving openings in the soaling about 1 inch wide. I would depend more on the close fit of the wheel to the breast ark for keeping the water on the wheel than on the form of the buckets; yet they should be shaped to receive and part with the water freely. Wood buckets discharge the impelling water on the breast ark sooner than iron buckets, which is a loss of power, so that there is an advantage in making them of iron, as they can be curved to any shape that may be required.

IMPROVED BREAST-WHEEL.

WHEN we want to perform the largest amount of work by a given quantity of water, the iron curved buckets and a slow-speeded wheel are the best arrangements. In this case the power is taken from the loaded side of the wheel by means of internal or external spur-gear bolted to the shrouding, and thereby transmitting a quick speed to the machinery to be driven. The slower the wheel moves and the higher it receives the water, the more the buckets may be sloped, so as to retain the water in them to the proper place of discharge. Another advantage is gained in this class of wheel by conducting the water into the buckets over the sluice from the upper surface of the water in the head race, which gains power in proportion to the depth of the head of water over the wheel. As experience proves that 1 foot of fall will give as much power as 2 feet of head, it is thereby evident that no more head should be adopted than is sufficient to give a proper velocity to the wheel. The speed at the circumference commonly in use is about $4\frac{1}{2}$ feet per second, and the speed of the entering water about $6\frac{1}{2}$ feet per second, for wheels of this class; and, if well constructed, there is no other kind of vertical wheel will do the same amount of work with a like quantity of water. The capacity of the buckets is made by considering the quantity of water that is to supply them. For example: suppose four buckets to pass the sluice in one second, and that 12 cubic feet of water were given to supply them during the same time, then we should have 3 cubic feet for each bucket; but they are made to hold double that quantity, or 6 cubic feet.

Or if we suppose 12 cubic feet of water to be delivered on the wheel per second, and the velocity of its circumference to be 5 feet per second, and space of its buckets 15 inches: then $\frac{12}{5} = 2\cdot4$ feet sectional area for each foot of circumference; but we want the area for 15 inches, consequently the area required will be 6 for $\frac{2 \cdot 4}{4} = \cdot 6$ and $2 \cdot 4 + \cdot 6 = 3$ cubic feet for each bucket; but they must contain twice that quantity = 6 cubic feet. By dividing the depth of the shrouding into the cubic feet for each bucket, we get the length of each bucket; or by dividing the length of a bucket into the cubic feet we get the depth of the shrouding.

EXAMPLE. Given 6 cubic feet, and depth of shrouding 1·5 foot; required the length of buckets, $\frac{6}{1 \cdot 5} = 4$ feet length of buckets, and $\frac{6}{4} = 1 \cdot 5$ depth of shrouding.

In the course of my practice, making these kind of wheels, I have found that a 10-inch shrouding would be a good proportion for 30-horse power on a waterfall of 40 feet, 10¾ for 30-horse power, &c., &c. And assuming 10 feet for the lowest fall, and shrouding 14½ inches deep, I have deduced the following series of proportional numbers, corresponding to the different falls and horse-powers; and I would make the space of the buckets measured on the periphery about 13 inches for narrow shrouding, and 18 for deep shrouding; or 7 added to half the depth of the shroud will give the space. Example: given the depth of the shrouding 12 inches, the space of the buckets would be 13 inches, for $6 + 7 = 13$, or if the shroud was 22 inches deep, the space by this rule would be 18 inches.

TABLE of the DEPTH of SHROUDING.

30-horse power.		40-horse power.		50-horse power.	
Height of Fall in feet.	Depth of Shrouding in inches.	Height of Fall in feet.	Depth of Shrouding in inches.	Height of Fall in feet.	Depth of Shrouding in inches.
40	10	40	10¾	40	11¼
35	10¾	35	12	35	12⅞
30	11½	30	13¼	30	14½
25	12¼	25	14½	25	16⅛
20	13	20	15¾	20	17¾
15	13¾	15	17	15	19⅜
10	14½	10	18¼	10	21

OVERSHOT WATER-WHEELS.

THE overshot wheel has the water let into it at a little distance from the top; and when economy of the propelling water is not a particular object, the circumference of this wheel may be driven at a speed of 8½ feet per second under a waterfall of 30 feet. This quick speed will cause the wheel to be much lighter and cheaper than the slow-motion wheel, which will in some degree compensate for the loss of power by its adoption. If the buckets be made of iron, having the proper curvature, and the wheel well proportioned and balanced, 12 cubic feet of water per second will give one-horse power on every foot of fall. I shall now propose rules for the dimensions of these wheels, and afterwards endeavour to illustrate the construction of the buckets for each kind of wheels by such diagrams and examples as I have found to be agreeable to practice.

DEFINITIONS AND FORMULA.

Head of Discharge is the perpendicular height from the surface of the water on the head race to the centre of the sluice opening.

Head of Contact is the distance in feet from the surface of the water until it meets the wheel.

Fall is the perpendicular distance from where the water meets the wheel to the surface of the water in the tail race at the time the wheel is working.

Effective Fall is half the head of contact added to the fall, and this sum multiplied by the cubic feet per second and divided by 12 will be the number of horse-power of an overshot or breast wheel at about 73 per cent.; or by dividing any other per cent. into 8·75 will obtain a corresponding factor. For example: the factor for 33 per cent. is 27, for $\dfrac{8 \cdot 75}{33} = 27$.

OVERSHOT WATER-WHEELS.

FORMULA FOR WHEELS OF QUICK SPEED EQUAL TO 10-HORSE POWER.

H The total height of the waterfall in feet.

D The exterior diameter of the wheel $= \overset{c}{1\cdot 08} \times H.$

V The velocity of its circumference per second $= \overset{c}{2\cdot 3} \sqrt[3]{D}.$

d The depth of the shrouding $= \dfrac{\overset{c}{6\cdot 8}}{V}.$

r Radius of the wheel.

N The number of buckets $= r \times 4 + r.$

L The length of the buckets, for 10-horse power $= \dfrac{\overset{c}{380}}{d \times V \times H}.$

h The height of the head of contact $= \dfrac{{}^2V}{21}.$

R The revolutions per minute $= \dfrac{V \times 60}{D \times \dfrac{22}{7}}.$

Note.—c is placed over constant numbers.

EXAMPLE OF THE FOREGOING RULES.

Suppose a waterfall of 37·1 feet were given to find the proportions of an overshot wheel to be equal to 10-horse power. A diagram of this wheel is shown at Fig. 3, Plate XII.

$\overset{c}{1\cdot 08} \times 37\cdot 1 = 40$ feet, diameter of wheel.

$\overset{c}{2\cdot 3} \sqrt[3]{40} = 8\cdot 2$ feet, velocity of periphery.

$\dfrac{6\cdot 8}{8\cdot 2} = \cdot 829$ foot, depth of shrouding.

$\dfrac{40}{2} = 20$ feet, radius of wheel.

$20 \times 4 + 20 = 110$ feet, number of buckets.

$\dfrac{380}{\cdot 92 \times 8\cdot 2 \times 37\cdot 1} = 1\cdot 35$ foot, length of buckets.

$\dfrac{{}^2 8\cdot 2}{21} = 3\cdot 2$ feet, head of contact.

$\dfrac{8\cdot 2 \times 60}{40 \times \dfrac{22}{7}} = 3\cdot 9$ revolutions of the wheel per minute.

FORMATION OF WATER-WHEEL BUCKETS.

ANOTHER EXAMPLE.

Given a waterfall 14·9 feet high; required the speed and proportions of a water-wheel to be equal to 10-horse power. See Fig. 2, Plate XII., for the diagram of this wheel.

$$1 \overset{c}{\cdot} 08 \times 14 \cdot 9 = 16 \text{ feet, diameter of the wheel.}$$

$$2 \overset{c}{\cdot} 3 \sqrt[3]{16} = 5 \cdot 79 \text{ feet, velocity of circumference.}$$

$$\frac{6 \cdot 8}{5 \cdot 79} = 1 \cdot 174 \text{ foot, depth of shrouding.}$$

$$\frac{16}{2} = 8 \text{ feet, radius of wheel.}$$

$$8 \times 4 + 8 = 40 \text{ feet, number of buckets.}$$

$$\frac{\overset{c}{380}}{1 \cdot 31 \times 5 \cdot 79 \times 14 \cdot 9} = 3 \cdot 36 \text{ feet, length of buckets.}$$

$$\frac{5 \cdot 79}{21} = 1 \cdot 59 \text{ foot, head of contact.}$$

$$\frac{5 \cdot 79 \times 60}{16 \times \frac{22}{7}} = 6 \cdot 91 \text{ revolutions of the wheel per minute.}$$

Note.—The power of these wheels may be increased or diminished by lengthening or shortening the buckets in the above proportion, observing that 9 inches should be added to the length of the buckets in each wheel.

FORMATION OF WATER-WHEEL BUCKETS.

PLATE XI. Fig. 4 is a segment of a water-wheel 17 feet in diameter, with 32 wood buckets 2 feet 7 inches deep. These buckets are inclined up the stream at an angle of 10 degrees from the radius A H, and the mean direction of the water entering them is at an angle of about 30 degrees, with a tangent D E to the point where the water meets the wheel; and F E is the mean direction of the water, H A the radius line, and C B A the angle to give to the buckets. By referring to the Table for proportioning undershot water-wheels we shall find that a 17-feet water-wheel is for a waterfall of 6·6 feet; head of contact, 5 feet; velocity of circumference, 9·49 feet per second; 32 buckets 2 feet 1 inch deep for iron buckets, but they are made 2 feet 7 inches deep for wood buckets that are not curved.

Iron Buckets.

Further illustration of a 17-feet undershot water-wheel, with iron buckets, is delineated in Plate XI., Fig. 3. It has 32 buckets, 2 feet 1 inch deep, but they are inclined up the stream at an angle of 20 degrees from the radius line H A, and the radius to form their curves is one-eighth the diameter of the wheel. A B C is the angle to give to the buckets; and the centre O, while describing the curve B C G, is so placed that the segment will exactly cross the points at the inner and outer periphery of the shrouding at B C; F E, the mean direction of the water entering the wheel.

DESCRIPTION OF WATER-WHEELS.

Plate XII. is diagrams of four water-wheel shroudings, showing the shape and position of both wood and iron buckets, and the mean direction of the water entering them. Fig. 1 is designed for 40-horse power on a waterfall of 10 feet. The wheel, 17 feet diameter, to have a speed at the external periphery of 4½ feet per second, and shrouding 18 inches deep, with 40 buckets. Fig. 2 is another segment of a wheel, 16 feet diameter, to have a speed of 5·79 feet at the external periphery, and to be driven by a waterfall of 14·9 feet (see page 53, where all the necessary proportions of this wheel are calculated). Fig. 3 is a segment of a wheel, 40 feet diameter, to have a speed at the periphery of 8·2 feet per second. The other necessary proportions of this wheel are calculated at page 52. Fig. 4 is a second illustration of a wheel, 40 feet diameter, to have a speed at the outer circumference of 4½ feet per second, shroud 12 inches deep, with 112 buckets.

FORM OF WOOD BUCKETS.

Fig. 1. The depth of the shroud on the radius line A D is divided into five equal parts; two of these parts B D are taken for the inner portion of the buckets, and the position and depth of the other portion are found by making the angle C B E, which is an angle to the radius of 1 inch to each foot of the entire waterfall; and the point where the line E B intersects the periphery of the wheel at F will be the extremity of the outer portion of the bucket B F. This slope I gave to wheels working on waterfalls less than 20 feet, but I found that three-quarters of an inch to the foot would be sufficient slope for higher falls. The mean direction of the water entering the wheel is found by dividing the depth of the shroud at No. 5 into three equal parts J I H G, and the distance between the buckets K M into two equal parts; then a straight line is made from H to L, which line, when prolonged, will be the mean direction to give the water entering the wheel.

Buckets made of iron for wheels having either quick or slow speeds are formed from these straight lines; for example: at No. 6 the half-depth of the shroud is taken for a radius to form the curved part of the bucket; that is, where the first and third division on the shroud intersect the straight buckets at N O. This shape of buckets is for wheels that have a quick speed. The construction of curved buckets for slow speeds is delineated at No. 7, where twice the distance R F is taken for a radius to form the curve over three divisions on the shroud from F to R; and the inner portion of the curve R P is formed by a radius equal to half the depth of the shroud. An addition is put to the depth of the shrouding for compartments by which the air in the buckets makes its escape in the direction indicated by the arrows.

TABLE of the DIAMETER of WHEELS at the PITCH CIRCLE, from 11 to 300 Teeth.

Number of Teeth.	Pitch of the Teeth.									
	inch. 1¾.	inch. 1⅞.	inches. 2.	inches. 2⅛.	inches. 2¼.	inches. 2½.	inches. 2¾.	inches. 3.	inches. 3¼.	inches. 3½.
11	0 6¼	0 6⅝	0 7	0 7½	0 7⅞	0 8¾	0 9¾	0 10⅝	0 11⅜	1 0¼
12	0 6¾	0 7⅛	0 7⅝	0 8⅛	0 8⅝	0 9⅝	0 10⅝	0 11½	1 0⅜	1 1⅜
13	0 7⅜	0 7⅞	0 8⅜	0 8⅞	0 9⅜	0 10⅜	0 11½	1 0½	1 1⅜	1 2½
14	0 7⅞	0 8½	0 9	0 9½	0 10	0 11¼	1 0⅜	1 1½	1 2½	1 3⅝
15	0 8½	0 9	0 9⅝	0 10¼	0 10¾	1 0	1 1¼	1 2⅜	1 3½	1 4¾
16	0 9	0 9⅝	0 10¼	0 10⅞	0 11½	1 0¾	1 2	1 3⅜	1 4½	1 5⅞
17	0 9⅝	0 10¼	0 10⅞	0 11½	1 0¼	1 1½	1 2⅞	1 4¼	1 5½	1 6⅞
18	0 10	0 10¾	0 11½	1 0¼	1 0⅞	1 2⅜	1 3¾	1 5¼	1 6⅝	1 8
19	0 10⅝	0 11⅜	1 0⅛	1 0⅞	1 1⅝	1 3⅛	1 4⅝	1 6¼	1 7⅝	1 9⅛
20	0 11¼	1 0	1 0¾	1 1½	1 2⅜	1 4	1 5½	1 7⅛	1 8¾	1 10¼
21	0 11¾	1 0½	1 1½	1 2¼	1 3	1 4¾	1 6⅜	1 8⅛	1 9¾	1 11⅜
22	1 0⅜	1 1⅛	1 2	1 2⅞	1 3¾	1 5½	1 7¼	1 9	1 10¾	2 0½
23	1 0⅞	1 1¾	1 2⅝	1 3½	1 4½	1 6⅜	1 8	1 10	1 11¾	2 1⅝
24	1 1½	1 2⅜	1 3⅜	1 4¼	1 5¼	1 7⅛	1 9	1 10⅞	2 0⅞	2 2¾
25	1 2	1 2⅞	1 3⅞	1 4⅞	1 6	1 8	1 9⅞	1 11⅞	2 1⅞	2 3⅞
26	1 2½	1 3½	1 4½	1 5½	1 6⅝	1 8¾	1 10¾	2 0⅞	2 2⅞	2 4⅞
27	1 3	1 4⅛	1 5¼	1 6¼	1 7⅜	1 9½	1 11⅝	2 1¾	2 3⅞	2 6⅛
28	1 3⅝	1 4⅝	1 5¾	1 6⅞	1 8	1 10¼	2 0½	2 2¾	2 4⅞	2 7¼
29	1 4⅛	1 5⅜	1 6½	1 7⅝	1 8¾	1 11⅛	2 1⅜	2 3¾	2 6	2 8⅜
30	1 4½	1 6	1 7⅛	1 8¼	1 9½	2 0	2 2¼	2 4⅝	2 7	2 9⅜
31	1 5⅜	1 6½	1 7¾	1 9	1 10¼	2 0¾	2 3⅛	2 5⅝	2 8	2 10½
32	1 5⅞	1 7⅛	1 8⅜	1 9⅝	1 11	2 1½	2 4	2 6½	2 9⅛	2 11⅝
33	1 6½	1 7¾	1 9	1 10⅜	1 11⅝	2 2¼	2 4⅞	2 7½	2 10⅛	3 0¾
34	1 7	1 8⅜	1 9⅝	1 11	2 0⅜	2 3	2 5¾	2 8½	2 11⅛	3 1⅞
35	1 7½	1 9	1 10⅛	1 11¾	2 1	2 3⅞	2 6⅝	2 9½	3 0¼	3 3
36	1 8	1 9½	1 10⅞	2 0⅜	2 2	2 4⅝	2 7½	2 10⅜	3 1¼	3 4⅛
37	1 8⅝	1 10	1 11½	2 1	2 2½	2 5½	2 8⅜	2 11⅜	3 2¼	3 5¼
38	1 9¼	1 10¾	2 0¼	2 1¾	2 3¼	2 6¼	2 9¼	3 0¼	3 3⅜	3 6⅜
39	1 9¾	1 11⅜	2 0⅞	2 2⅜	2 4	2 7	2 10⅛	3 1¼	3 4⅜	3 7⅝
40	1 10⅜	1 11⅞	2 1½	2 3	2 4⅝	2 7⅞	2 10¾	3 2¼	3 5⅜	3 8½
41	1 10⅞	2 0½	2 2⅛	2 3¾	2 5⅜	2 8⅝	2 11⅞	3 3⅛	3 6⅜	3 9⅝
42	1 11½	2 1	2 2¾	2 4½	2 6	2 9⅜	3 0¾	3 4⅛	3 7⅜	3 10¾
43	2 0	2 1⅝	2 3⅜	2 5	2 6¾	2 10¼	3 1⅝	3 5	3 8⅜	4 0
44	2 0½	2 2¼	2 4	2 5¾	2 7½	2 11	3 2½	3 6	3 9½	4 1
45	2 1	2 2⅞	2 4⅝	2 6½	2 8¼	2 11¾	3 3⅜	3 7	3 10½	4 2¼
46	2 1⅝	2 3½	2 5¼	2 7⅛	2 9	3 0⅝	3 4¼	3 7⅞	3 11½	4 3¼
47	2 1⅞	2 4	2 6	2 7⅞	2 9⅝	3 1½	3 5⅛	3 8⅞	4 0⅝	4 4⅜
48	2 2¾	2 4⅝	2 6½	2 8½	2 10⅜	3 2¼	3 6	3 9⅞	4 1⅝	4 5½
49	2 3⅜	2 5¼	2 7¼	2 9⅛	2 11	3 3	3 6⅞	3 10⅞	4 2⅝	4 6½
50	2 3⅞	2 5⅞	2 7⅞	2 9¾	2 11¾	3 3¾	3 7¾	3 11¾	4 3¾	4 7¾
51	2 4½	2 6½	2 8½	2 10½	3 0½	3 4½	3 8⅝	4 0¾	4 4¾	4 8⅞
52	2 4⅞	2 7⅛	2 9⅛	2 11⅛	3 1¼	3 5⅜	3 9½	4 1⅝	4 5¾	4 10
53	2 5⅜	2 7⅝	2 9¾	2 11⅞	3 2	3 6¼	3 10½	4 2⅝	4 6⅞	4 11
54	2 6	2 8¼	2 10⅜	3 0½	3 2⅝	3 7	3 11	4 3½	4 7⅞	5 0⅛
55	2 6⅝	2 8⅞	2 11	3 1¼	3 3½	3 7¾	4 0⅛	4 4½	4 8⅞	5 1¼
56	2 7⅛	2 9⅜	2 11⅝	3 1⅞	3 4⅛	3 8¼	4 1	4 5½	4 9⅞	5 2⅜
57	2 7¾	2 10	3 0¼	3 2½	3 4⅞	3 9⅜	4 1⅞	4 6⅜	4 10⅞	5 3½

Table of the Diameter of Wheels at the Pitch Circle—*continued.*

Number of Teeth.	Pitch of the Teeth.									
	inch. 1¾.	inch. 1⅞.	inches. 2.	inches. 2⅛.	inches. 2¼.	inches. 2½.	inches. 2¾.	inches. 3.	inches. 3¼.	inches. 3½.
58	2 8¼	2 10⅝	3 0⅞	3 3¼	3 5½	3 10⅛	4 2¾	4 7⅜	5 0	5 4⅝
59	2 8⅞	2 11¼	3 1½	3 4	3 6¼	3 11⅛	4 3⅝	4 8⅜	5 1	5 5¾
60	2 9⅜	2 11¾	3 2¼	3 4⅝	3 7	3 11¾	4 4½	4 9¼	5 2	5 6⅞
61	2 10	3 0⅜	3 2⅞	3 5¼	3 7¾	4 0½	4 5¾	4 10¼	5 3⅛	5 8
62	2 10½	3 1	3 3½	3 6	3 8½	4 1⅜	4 6¼	4 11¼	5 4⅛	5 9
63	2 11	3 1⅝	3 4⅛	3 6⅝	3 9⅛	4 2⅛	4 7⅛	5 0⅛	5 5⅛	5 10⅛
64	2 11⅝	3 2¼	3 4¾	3 7¼	3 9⅞	4 3	4 8	5 1⅛	5 6¼	5 11⅜
65	3 0¼	3 2⅞	3 5⅜	3 8	3 10½	4 3¾	4 8⅞	5 2	5 7¼	6 0⅜
66	3 0⅞	3 3⅜	3 6	3 8⅝	3 11¼	4 4½	4 9¾	5 3	5 8¼	6 1½
67	3 1⅜	3 4	3 6½	3 9⅜	4 0	4 5⅜	4 10⅝	5 4	5 9⅜	6 2⅝
68	3 1¾	3 4⅝	3 7¼	3 10	4 0¾	4 6⅛	4 11½	5 5	5 10⅜	6 3¾
69	3 2⅜	3 5⅛	3 7⅞	3 10⅝	4 1½	4 7	5 0⅜	5 6	5 11⅜	6 4⅞
70	3 3	3 5¾	3 8½	3 11⅜	4 2⅛	4 7¾	5 1¼	5 6⅞	6 0½	6 6
71	3 3½	3 6⅜	3 9¼	4 0	4 2⅞	4 8½	5 2⅛	5 7¾	6 1½	6 7
72	3 4⅛	3 6¾	3 9⅞	4 0¾	4 3½	4 9¼	5 3	5 8¾	6 2½	6 8¼
73	3 4⅝	3 7½	3 10½	4 1⅜	4 4¼	4 10	5 3⅞	5 9¾	6 3½	6 9⅜
74	3 5¼	3 7⅞	3 11⅛	4 2	4 5	4 10⅞	5 4¾	5 10⅝	6 4½	6 10⅝
75	3 5¾	3 8¾	3 11¾	4 2¾	4 5¾	4 11¾	5 5⅜	5 11⅝	6 5½	6 11½
76	3 6⅜	3 9⅜	4 0⅜	4 3½	4 6½	5 0½	5 6½	6 0½	6 6⅝	7 0⅝
77	3 6⅞	3 9⅞	4 1	4 4	4 7⅛	5 1¼	5 7⅜	6 1½	6 7⅝	7 1¼
78	3 7½	3 10½	4 1⅝	4 4¾	4 7⅞	5 2	5 8¼	6 2½	6 8⅝	7 2⅞
79	3 8	3 11⅛	4 2¼	4 5½	4 8½	5 2⅞	5 9⅛	6 3½	6 9¾	7 4
80	3 8½	3 11¾	4 3	4 6⅛	4 9¼	5 3⅝	5 10	6 4⅜	6 10¾	7 5⅛
81	3 9⅛	4 0⅜	4 3½	4 6⅞	4 10	5 4½	5 10⅞	6 5⅜	6 11⅞	7 6¼
82	3 9⅝	4 0⅞	4 4¼	4 7½	4 10¾	5 5¼	5 11¾	6 6⅜	7 0⅞	7 7⅜
83	3 10¼	4 1½	4 4⅞	4 8⅛	4 11½	5 6	6 0⅝	6 7¼	7 1⅞	7 8⅝
84	3 10¾	4 2⅛	4 5½	4 8⅞	5 0⅛	5 6⅞	6 1½	6 8⅛	7 2⅞	7 9½
85	3 11¼	4 2¾	4 6⅛	4 9½	5 0⅞	5 7⅝	6 2⅜	6 9⅛	7 3⅞	7 10⅝
86	3 11⅞	4 3¼	4 6¾	4 10⅛	5 1⅝	5 8½	6 3¼	6 10½	7 5	7 11¾
87	4 0½	4 3⅞	4 7⅜	4 10⅞	5 2¼	5 9¼	6 4⅛	6 11	7 6	8 0⅞
88	4 1	4 4½	4 8	4 11½	5 3	5 10	6 5	7 0	7 7	8 2
89	4 1½	4 5⅛	4 8⅝	5 0¼	5 3¼	5 10¾	6 5⅞	7 1	7 8	8 3⅛
90	4 2⅛	4 5¾	4 9¼	5 0⅞	5 4¼	5 11⅝	6 6¾	7 2	7 9⅛	8 4¼
91	4 2¾	4 6¼	4 9⅞	5 1½	5 5⅛	6 0⅜	6 7⅝	7 2⅞	7 10⅛	8 5⅜
92	4 3¼	4 7	4 10½	5 2⅛	5 5⅞	6 1	6 8½	7 3⅞	7 11⅛	8 6½
93	4 3⅞	4 7½	4 11¼	5 2⅞	5 6⅝	6 2	6 9⅜	7 4⅞	8 0¼	8 7⅝
94	4 4⅜	4 8⅛	4 11¾	5 3½	5 7⅜	6 2¾	6 10¼	7 5¾	8 1¼	8 8¾
95	4 4⅞	4 8¾	5 0½	5 4¼	5 8	6 3½	6 11⅛	7 6¾	8 2¼	8 9⅞
96	4 5½	4 9⅜	5 1⅛	5 5	5 8¾	6 4⅜	7 0	7 7⅝	8 3⅜	8 10⅞
97	4 6	4 10	5 1¾	5 5⅝	5 9½	6 5¼	7 0⅞	7 8⅝	8 4⅜	9 0
98	4 6½	4 10½	5 2⅜	5 6¼	5 10⅛	6 6	7 1¾	7 9½	8 5⅜	9 1⅛
99	4 7⅛	4 11	5 3	5 7	5 11	6 6¾	7 2⅝	7 10½	8 6½	9 2¼
100	4 7¾	4 11⅝	5 3⅝	5 7⅝	5 11⅝	6 7½	7 3⅝	7 11½	8 7½	9 3⅜
101	4 8¼	5 0¼	5 4¼	5 8¼	6 0¼	6 8⅜	7 4⅜	8 0½	8 8½	9 4½
102	4 8⅞	5 1	5 5	5 9	6 1	6 9⅛	7 5¼	8 1⅜	8 9½	9 5⅝
103	4 9⅜	5 1¼	5 5½	5 9⅝	6 1½	6 10	7 6⅛	8 2¼	8 10½	9 6¾
104	4 10	5 1¾	5 6¼	5 10¼	6 2¼	6 10¾	7 7	8 3¼	8 11½	9 7⅞

TABLE of the DIAMETER of WHEELS at the PITCH CIRCLE—*continued.*

Number of Teeth.	Pitch of the Teeth.									
	inch. 1¾.	inch. 1⅞.	inches. 2.	inches. 2⅛.	inches. 2¼.	inches. 2½.	inches. 2¾.	inches. 3.	inches. 3¼.	inches. 3½.
105	4 10½	5 2⅜	5 6¾	5 11	6 3	6 11½	7 7⅞	8 4¼	9 0⅝	9 8⅞
106	4 11	5 2⅞	5 7½	5 11⅝	6 3⅝	7 0¼	7 8¾	8 5⅛	9 1⅝	9 10
107	4 11½	5 3½	5 8½	6 0⅜	6 4¼	7 1⅛	7 9⅝	8 6⅛	9 2⅝	9 11¼
108	5 0⅛	5 4⅛	5 8¾	6 1	6 5	7 2	7 10½	8 7⅛	9 3¾	10 0⅜
109	5 0¾	5 4¾	5 9⅜	6 1¾	6 5¾	7 2¾	7 11⅜	8 8	9 4¾	10 1⅜
110	5 1¼	5 5¼	5 10	6 2⅜	6 6½	7 3½	8 0¼	8 9	9 5¾	10 2⅜
111	5 1⅞	5 5⅞	5 10⅝	6 3	6 7⅛	7 4¼	8 1⅛	8 10	9 6⅞	10 3⅝
112	5 2⅜	5 6½	5 11¼	6 3¾	6 8	7 5⅛	8 2	8 10⅞	9 7⅞	10 4¾
113	5 3	5 7	6 0	6 4⅜	6 8⅝	7 6	8 3	8 11⅞	9 8⅞	10 6
114	5 3½	5 7⅝	6 0½	6 5⅛	6 9¼	7 6¾	8 3¾	9 0¾	9 9⅞	10 7
115	5 4	5 8¼	6 1¼	6 5¾	6 10	7 7½	8 4⅝	9 1¾	9 10⅞	10 8⅛
116	5 4⅝	5 8⅞	6 1¾	6 6½	6 10¾	7 8¼	8 5½	9 2¾	10 0	10 9¼
117	5 5⅛	5 9½	6 2½	6 7⅛	6 11½	7 9⅛	8 6⅜	9 3⅝	10 1	10 10⅜
118	5 5¾	5 10	6 3⅛	6 7¾	7 0⅛	7 10	8 7¼	9 4⅝	10 2	10 11⅝
119	5 6¼	5 10⅝	6 3¾	6 8½	7 1	7 10¾	8 8⅛	9 5½	10 3⅛	11 0½
120	5 6¾	5 11¼	6 4⅜	6 9⅛	7 1⅝	7 11½	8 9	9 6½	10 4⅛	11 1⅝
121	5 7⅜	5 11⅞	6 5	6 9¾	7 2¼	8 0¼	8 9⅞	9 7½	10 5⅛	11 2¾
122	5 8	6 0½	6 5⅝	6 10½	7 3	8 1	8 10¾	9 8½	10 6¼	11 3⅞
123	5 8½	6 1	6 6¼	6 11⅛	7 3¾	8 1⅞	8 11⅝	9 9⅜	10 7¼	11 5
124	5 9	6 1⅝	6 7	6 11⅞	7 4½	8 2⅝	9 0½	9 10⅜	10 8¼	11 6⅛
125	5 9⅝	6 2¼	6 7½	7 0½	7 5⅛	8 3½	9 1⅜	9 11¼	10 9⅜	11 7¼
126	5 10⅛	6 2¾	6 8¼	7 1¼	7 6	8 4¼	9 2¼	10 0¼	10 10⅜	11 8⅜
127	5 10¾	6 3⅜	6 8¾	7 2	7 6⅝	8 5	9 3⅛	10 1¼	10 11⅜	11 9½
128	5 11¼	6 4	6 9½	7 2½	7 7¼	8 5⅞	9 4	10 2¼	11 0½	11 10½
129	5 11⅞	6 4½	6 10⅛	7 3¼	7 8	8 6⅝	9 4⅞	10 3⅛	11 1½	11 11¾
130	6 0⅜	6 5⅛	6 10¾	7 4	7 8¾	8 7½	9 5¾	10 4⅛	11 2½	12 0⅞
131	6 1	6 5¾	6 11⅜	7 4½	7 9½	8 8¼	9 6⅝	10 5	11 3½	12 2
132	6 1½	6 6⅜	7 0	7 5⅛	7 10⅛	8 9	9 7½	10 6	11 4⅝	12 3
133	6 2	6 7	7 0⅝	7 6	7 10⅞	8 9¾	9 8⅜	10 7	11 5⅝	12 4⅛
134	6 2⅝	6 7½	7 1¼	7 6⅝	7 11½	8 10⅝	9 9¼	10 8	11 6⅝	12 5¼
135	6 3⅛	6 8½	7 2	7 7¼	8 0¼	8 11⅜	9 10⅛	10 8⅞	11 7⅝	12 6¾
136	6 3¾	6 8¾	7 2½	7 8	8 1	9 0¼	9 11	10 9¾	11 8⅝	12 7½
137	6 4¼	6 9¼	7 3¼	7 8⅝	8 1¾	9 1	10 0	10 10¾	11 9¾	12 8⅝
138	6 4¾	6 10	7 3¾	7 9¼	8 2½	9 1¾	10 0⅞	10 11¾	11 10¾	12 9¾
139	6 5½	6 10½	7 4½	7 10	8 3⅛	9 2⅝	10 1⅝	11 0⅝	11 11¾	12 10⅞
140	6 6	6 11⅛	7 5⅛	7 10⅝	8 3⅞	9 3⅜	10 2½	11 1⅝	12 0⅞	13 0
141	6 6½	6 11¾	7 5¾	7 11⅜	8 4½	9 4¼	10 3⅜	11 2½	12 1⅞	13 1
142	6 7	7 0¼	7 6¾	8 0	8 5¼	9 5	10 4¼	11 3½	12 2⅞	13 2⅛
143	6 7⅝	7 0⅞	7 7	8 0¾	8 6	9 5¾	10 5⅛	11 4½	12 3⅞	13 3⅜
144	6 8¼	7 1½	7 7⅝	8 1⅜	8 6¾	9 6	10 6	11 5½	12 4⅞	13 4⅜
145	6 8¾	7 2	7 8¼	8 2	8 7½	9 7⅜	10 6⅞	11 6½	12 6	13 5½
146	6 9¼	7 2⅝	7 9	8 2¾	8 8½	9 8⅛	10 7¾	11 7⅜	12 7	13 6⅝
147	6 9⅞	7 3¼	7 9½	8 3⅜	8 8⅞	9 9	10 8⅝	11 8¼	12 8	13 7¾
148	6 10½	7 3⅞	7 10¼	8 4⅛	8 9½	9 9¾	10 9½	11 9¼	12 9⅛	13 8⅞
149	6 11	7 4½	7 10¾	8 4⅞	8 10¼	9 10½	10 10⅜	11 10¼	12 10⅛	13 10
150	6 11½	7 5	7 11½	8 5½	8 11	9 11⅜	10 11¼	11 11¼	12 11⅛	13 11⅛

TABLE of the DIAMETER of WHEELS at the PITCH CIRCLE—*continued.*

Number of Teeth.	Pitch of the Teeth.					
	inches. 2¼.	inches. 2½.	inches. 2¾.	inches. 3.	inches. 3¼.	inches. 3½.
151	9 0⅛	10 0⅛	11 0⅛	12 0⅛	13 0⅛	14 0¼
152	9 0⅞	10 0⅞	11 1	12 1⅛	13 1¼	14 1⅜
153	9 1⅝	10 1¾	11 1⅞	12 2	13 2¼	14 2⅜
154	9 2¼	10 2½	11 2⅞	12 3	13 3⅜	14 3½
155	9 3	10 3⅜	11 3⅝	12 4	13 4⅜	14 4⅝
156	9 3⅝	10 4⅛	11 4½	12 4⅞	13 5⅜	14 5¾
157	9 4⅜	10 4⅞	11 5⅝	12 5⅞	13 6⅜	14 6⅞
158	9 5⅛	10 5⅝	11 6⅜	12 6⅞	13 7⅜	14 8
159	9 5⅞	10 6½	11 7⅛	12 7⅞	13 8⅜	14 9⅛
160	9 6½	10 7⅜	11 8	12 8¾	13 9½	14 10¼
161	9 7⅜	10 8⅛	11 8⅞	12 9⅝	13 10½	14 11⅜
162	9 8	10 8⅞	11 9¾	12 10⅝	13 11½	15 0⅜
163	9 8¾	10 9⅝	11 10⅝	12 11⅝	14 0¾	15 1½
164	9 9⅜	10 10½	11 11½	13 0⅝	14 1⅝	15 2⅝
165	9 10⅛	10 11⅜	12 0¾	13 1½	14 2⅝	15 3¾
166	9 10⅞	11 0	12 1¼	13 2½	14 3¾	15 4⅞
167	9 11⅝	11 0⅞	12 2⅛	13 3⅜	14 4⅜	15 6
168	10 0⅜	11 1⅝	12 3	13 4⅜	14 5¾	15 7⅛
169	10 1	11 2⅜	12 3⅞	13 5⅜	14 6⅝	15 8¼
170	10 1¾	11 3¼	12 4¾	13 6⅜	14 7⅝	15 9⅝
171	10 2⅜	11 4	12 5⅝	13 7¼	14 8⅞	15 10½
172	10 3⅛	11 4⅞	12 6½	13 8⅛	14 9⅞	15 11⅝
173	10 3⅞	11 5¾	12 7⅜	13 9⅛	14 10⅞	16 0⅝
174	10 4½	11 6⅜	12 8¼	13 10⅛	15 0	16 1¾
175	10 5⅜	11 7¼	12 9⅛	13 11⅛	15 1⅜	16 2⅞
176	10 6	11 8	12 10	14 0	15 2⅝	16 4
177	10 6¾	11 8⅞	12 10⅞	14 1	15 3⅛	16 5⅛
178	10 7⅝	11 9⅝	12 11⅞	14 1⅞	15 4¼	16 6¾
179	10 8⅛	11 10⅜	13 0⅝	14 2⅞	15 5⅛	16 7⅜
180	10 8⅞	11 11¼	13 1½	14 3⅞	15 6¼	16 8½
181	10 9⅝	12 0	13 2⅝	14 4⅞	15 7¼	16 9⅝
182	10 10⅜	12 0⅞	13 3⅜	14 5¾	15 8¼	16 10¾
183	10 11	12 1⅝	13 4⅛	14 6¾	15 9⅜	16 11⅞
184	10 11¾	12 2⅜	13 5	14 7⅝	15 10⅜	17 0⅞
185	11 0⅜	12 3⅛	13 5⅞	14 8⅝	15 11⅜	17 2⅛
186	11 1⅛	12 4	13 6⅞	14 9⅝	16 0⅜	17 3¼
187	11 1⅞	12 4⅞	13 7⅝	14 10½	16 1⅜	17 4⅜
188	11 2⅝	12 5½	13 8½	14 11½	16 2⅜	17 5⅜
189	11 3⅜	12 6⅜	13 9⅜	15 0⅜	16 3½	17 6½
190	11 4	12 7⅛	13 10⅜	15 1⅜	16 4½	17 7⅝
191	11 4¾	12 7⅞	13 11⅛	15 2⅜	16 5½	17 8¾
192	11 5½	12 8⅜	14 0	15 3⅜	16 6⅝	17 9⅞
193	11 6⅛	12 9½	14 0⅞	15 4¼	16 7⅝	17 11
194	11 6⅞	12 10⅜	14 1⅞	15 5¼	16 8⅝	18 0⅛
195	11 7⅝	12 11⅛	14 2⅝	15 6⅛	16 9¾	18 1¼

TABLE of the DIAMETER of WHEELS at the PITCH CIRCLE—*continued.*

Number of Teeth.	Pitch of the Teeth.					
	inches. 2¼.	inches. 2½.	inches. 2¾.	inches. 3.	inches. 3¼.	inches. 3½.
196	11 8⅜	12 11⅞	14 3½	15 7⅛	16 10¾	18 2⅜
197	11 9	13 0¾	14 4⅜	15 8⅛	16 11¾	18 3½
198	11 9⅞	13 1½	14 5⅜	15 9	17 0⅞	18 4½
199	11 10½	13 2⅜	14 6⅛	15 10	17 1⅞	18 5⅝
200	11 11⅛	13 3⅛	14 7	15 10⅞	17 2⅞	18 6¾
201	11 11⅞	13 3⅞	14 7⅞	15 11⅞	17 3⅞	18 7⅞
202	12 0⅝	13 4¾	14 8⅞	16 0⅞	17 4⅞	18 9
203	12 1⅜	13 5⅜	14 9⅝	16 1⅞	17 6	18 10
204	12 2	13 6¼	14 10½	16 2⅞	17 7	18 11⅛
205	12 2⅞	13 7⅛	14 11⅜	16 3¾	17 8	19 0¼
206	12 3½	13 7⅞	15 0⅜	16 4⅝	17 9⅛	19 1⅜
207	12 4¼	13 8⅝	15 1⅛	16 5⅝	17 10⅛	19 2½
208	12 4⅞	13 9⅜	15 2	16 6⅝	17 11⅛	19 3⅝
209	12 5⅝	13 10¼	15 2⅞	16 7½	18 0¼	19 4¾
210	12 6⅜	13 11⅛	15 3⅞	16 8½	18 1¼	19 5⅞
211	12 7⅛	13 11⅞	15 4⅝	16 9⅜	18 2¼	19 7
212	12 7⅞	14 0⅝	15 5½	16 10⅜	18 3⅛	19 8⅛
213	12 8½	14 1⅜	15 6⅜	16 11⅜	18 4⅜	19 9¼
214	12 9¼	14 2⅛	15 7⅜	17 0⅜	18 5⅞	19 10⅜
215	12 9⅞	14 3	15 8⅛	17 1⅜	18 6⅝	19 11½
216	12 10⅝	14 3⅞	15 9	17 2¼	18 7⅝	20 0⅝
217	12 11⅜	14 4⅝	15 9⅞	17 3⅛	18 8⅝	20 1¾
218	13 0	14 5⅜	15 10⅞	17 4⅛	18 9½	20 2⅞
219	13 0⅞	14 6¼	15 11⅝	17 5⅛	18 10½	20 4
220	13 1½	14 7	16 0½	17 6	18 11½	20 5⅛
221	13 2¼	14 7⅞	16 1⅜	17 7	19 0⅝	20 6¼
222	13 2⅞	14 8⅝	16 2⅜	17 7⅞	19 1⅝	20 7⅜
223	13 3⅝	14 9⅜	16 3⅛	17 8⅞	19 2⅞	20 8½
224	13 4⅜	14 10¼	16 4	17 9⅞	19 3¾	20 9⅝
225	13 5⅛	14 11	16 4⅞	17 10⅞	19 4¾	20 10¾
226	13 5⅞	14 11⅞	16 5⅞	17 11⅞	19 5¾	20 11⅞
227	13 6½	15 0⅝	16 6⅝	18 0¾	19 6⅞	21 0⅞
228	13 7¼	15 1⅜	16 7½	18 1⅝	19 7⅞	21 2
229	13 8	15 2⅛	16 8⅜	18 2⅝	19 8⅞	21 3⅛
230	13 8⅝	15 3	16 9⅜	18 3⅝	19 9⅞	21 4¼
231	13 9⅜	15 3⅞	16 10⅛	18 4½	19 10⅞	21 5⅜
232	13 10⅛	15 4⅝	16 11	18 5½	20 0	21 6⅜
233	13 10⅞	15 5⅜	16 11⅞	18 6⅜	20 1	21 7½
234	13 11½	15 6⅛	17 0⅞	18 7⅜	20 2	21 8⅝
235	14 0¼	15 7	17 1⅝	18 8⅜	20 3	21 9⅞
236	14 1	15 7⅞	17 2½	18 9⅜	20 4⅛	21 10⅞
237	14 1⅝	15 8½	17 3⅜	18 10⅜	20 5⅛	22 0
238	14 2⅜	15 9⅜	17 4⅜	18 11¼	20 6¼	22 7⅛
239	14 3⅛	15 10⅛	17 5⅛	19 0⅛	20 7¼	22 2¼
240	14 3⅞	15 10⅞	17 6	19 1⅛	20 8¼	22 3⅝

TABLE of the DIAMETER of WHEELS at the PITCH CIRCLE—*continued.*

Number of Teeth.	Pitch of the Teeth.					
	inches. 2¼.	inches. 2½.	inches. 2¾.	inches. 3.	inches. 3¼.	inches. 3½.
241	14 4½	15 11⅞	17 6⅞	19 2⅞	20 9⅜	22 4⅜
242	14 5¼	16 0½	17 7⅝	19 3⅞	20 10⅜	22 5⅝
243	14 6	16 1⅜	17 8⅝	19 4⅝	20 11⅜	22 6⅝
244	14 6¾	16 2⅛	17 9½	19 5½	21 0⅜	22 7⅞
245	14 7⅜	16 2⅞	17 10¾	19 6⅜	21 1⅝	22 8⅞
246	14 8⅛	16 3¾	17 11⅜	19 7⅜	21 2⅝	22 10
247	14 8⅞	16 4½	18 0⅛	19 8⅛	21 3½	22 11⅛
248	14 9⅝	16 5⅛	18 1	19 9	21 4½	23 0¼
249	14 10⅜	16 6⅛	18 1⅞	19 9⅞	21 5½	23 1⅛
250	14 11	16 6⅞	18 2⅞	19 10⅞	21 6⅝	23 2⅛
251	14 11¾	16 7⅝	18 3⅝	19 11⅝	21 7⅝	23 3⅝
252	15 0⅜	16 8½	18 4½	20 0⅝	21 8⅝	23 4¾
253	15 1⅛	16 9¼	18 5⅜	20 1½	21 9⅝	23 5⅞
254	15 1⅞	16 10⅛	18 6⅜	20 2½	21 10¾	23 6⅞
255	15 2⅝	16 10⅞	18 7⅛	20 3½	21 11¾	23 8
256	15 3⅜	16 11⅞	18 8	20 4⅜	22 0⅞	23 9¼
257	15 4	17 0½	18 8⅞	20 5⅜	22 1⅞	23 10⅜
258	15 4¾	17 1⅜	18 9⅞	20 6⅜	22 2⅞	23 11⅜
259	15 5⅜	17 2⅛	18 10¼	20 7⅜	22 3⅞	24 0½
260	15 6¼	17 2⅞	18 11½	20 8¼	22 4⅞	24 1⅝
261	15 6⅞	17 3⅝	19 0⅜	20 9⅛	22 6	24 2¾
262	15 7⅝	17 4⅜	19 1⅜	20 10⅛	22 7	24 3⅞
263	15 8⅜	17 5¼	19 2⅛	20 11⅛	22 8	24 5
264	15 9	17 6	19 3	21 0⅛	22 9⅛	24 6⅛
265	15 9¾	17 6⅞	19 3⅞	21 1	22 10⅛	24 7¼
266	15 10⅝	17 7⅝	19 4⅞	21 2	22 11⅛	24 8⅜
267	15 11⅛	17 8⅜	19 5⅝	21 2⅞	23 0¼	24 9⅜
268	15 11⅞	17 9¼	19 6½	21 3⅞	23 1¼	24 10½
269	16 0⅝	17 10	19 7⅜	21 4⅞	23 2¼	24 11⅝
270	16 1⅜	17 10⅞	19 8⅜	21 5⅞	23 3⅜	25 0⅞
271	16 2	17 11⅝	19 9⅜	21 6¾	23 4⅜	25 1⅞
272	16 2¾	18 0⅜	19 10	21 7⅝	23 5⅜	25 3
273	16 3½	18 1⅛	19 10⅞	21 8⅝	23 6⅜	25 4⅛
274	16 4⅛	18 2	19 11⅞	21 9⅝	23 7⅜	25 5¼
275	16 4⅞	18 2⅞	20 0⅝	21 10⅝	23 8⅜	25 6⅜
276	16 5⅜	18 3⅝	20 1½	21 11½	23 9½	25 7⅜
277	16 6⅜	18 4⅜	20 2⅜	22 0½	23 10½	25 8⅝
278	16 7	18 5¼	20 3⅜	22 1⅜	23 11½	25 9⅝
279	16 7⅞	18 6	20 4⅛	22 2⅜	24 0⅝	25 10⅞
280	16 8½	18 6⅞	20 5	22 3⅜	24 1⅝	25 11⅞
281	16 9¼	18 7⅝	20 5⅞	22 4⅜	24 2⅝	26 1
282	16 9⅞	18 8⅜	20 6⅞	22 5¼	24 3¾	26 2⅛
283	16 10⅝	18 9⅛	20 7⅝	22 6⅛	24 4¾	26 3¼
284	16 11⅜	18 9⅞	20 8½	22 7⅛	24 5¾	26 4⅜
285	17 0⅛	18 10¾	20 9⅜	22 8⅛	24 6⅞	26 5½

Table of the Diameter of Wheels at the Pitch Circle—*continued.*

Number of Teeth.	Pitch of the Teeth.					
	inches. 2¼.	inches. 2½.	inches. 2¾.	inches. 3.	inches. 3¼.	inches. 3½.
286	17 0⅞	18 11½	20 10⅜	22 9	24 7⅞	26 6⅝
287	17 1½	19 0⅜	20 11⅜	22 10	24 8⅞	26 7⅝
288	17 2¼	19 1⅛	21 0	22 11	24 9⅞	26 8⅞
289	17 2⅞	19 1⅞	21 0⅞	22 11⅞	24 10⅞	26 9⅞
290	17 3⅝	19 2⅜	21 1⅞	23 0⅞	25 0	26 11
291	17 4⅜	19 3½	21 2¾	23 1⅞	25 1	27 0⅛
292	17 5⅛	19 4½	21 3⅝	23 2⅞	25 2	27 1⅜
293	17 5⅞	19 5⅛	21 4⅜	23 3¾	25 3⅛	27 2⅜
294	17 6½	19 5⅞	21 5¼	23 4⅝	25 4⅜	27 3½
295	17 7¼	19 6⅜	21 6⅛	23 5⅝	25 5⅛	27 4⅝
296	17 7⅞	19 7½	21 7	23 6⅝	25 6¼	27 5¾
297	17 8⅝	19 8⅜	21 7⅞	23 7⅝	25 7¼	27 6⅞
298	17 9⅜	19 9⅛	21 8⅞	23 8½	25 8¼	27 7⅞
299	17 10⅛	19 9⅞	21 9⅝	23 9½	25 9⅜	27 9⅛
300	17 10⅞	19 10⅝	21 10⅝	23 10⅝	25 10⅜	27 10¼

Explanation of the Table on Wheels.

Suppose the diameter of a wheel was required to contain 118 teeth 2½-inch pitch. See 118 in the column headed "Number of Teeth," and on the same line, under 2½-inch pitch, is 7 feet 10 inches, the required diameter.

If we want to know the diameter of a wheel having a greater number of teeth than appears in the Table, we add together the diameters of two wheels, the sum of their teeth being equal to the number of teeth in the large wheel.

For example: we want to know the diameter of a wheel to contain 360 teeth 2½-inch pitch. By looking at the Table we see that a wheel having 300 teeth 2½-inch pitch is 19 feet 10⅝ inches diameter, and the diameter of a wheel with 60 teeth 2½-inch pitch is 3 feet 11¾ inches diameter; then 3 feet 11¾ inches + 19 feet 10⅝ inches = 23 feet 10⅜ inches, the required diameter.

The diameter in inches for any given pitch and number of teeth may be found by multiplying the number of teeth by the pitch, and the product by ·3183.

DIAMETER OF WHEELS.

EXAMPLE. Required the diameter of a wheel to contain 334 teeth 3½-inch pitch: $334 \times 3 \cdot 5 \times \cdot 3183 = 372$ inches, or 31 feet diameter.

A TABLE to FIND the STRENGTH of CAST-IRON TEETH of WHEELS at any GIVEN VELOCITY.

Numbers	19	14·75	11	7·5	4·75	3·5	2·75	2	1·5	1·2	1	·7	·5	·3
Pitch in inches	4	3½	3	2½	2	1¾	1½	1¼	1⅛	1	⅞	¾	⅝	½
Breadth	8	7	6	5	4	3½	3	2½	2¼	2	1¾	1½	1¼	1
Thickness	1·86	1·62	1·39	1·15	·92	·80	·68	·56	·50	·45	·38	·33	·31	·22

APPLICATION OF THIS TABLE.

Multiply the given velocity of the wheel at the pitch line per second by the number in the Table over the pitch; divide the product by 3, and the quotient will be the number of horse-power, the teeth having their breadth equal to twice the pitch. If the breadth be doubled the strength will be doubled; but if the pitch is doubled the strength is four times greater.

EXAMPLE. Given a velocity at the pitch line of 7 feet per second and 2-inch pitch; required the power this wheel is capable of transmitting, supposing the teeth to be 8 inches broad. We see that the number over 2-inch pitch is 4·75; then $\dfrac{4 \cdot 75 \times 7}{3} = 11 \cdot 08$ for 4 inches broad, and 22·16 horse-power will be the answer for 8 inches broad.

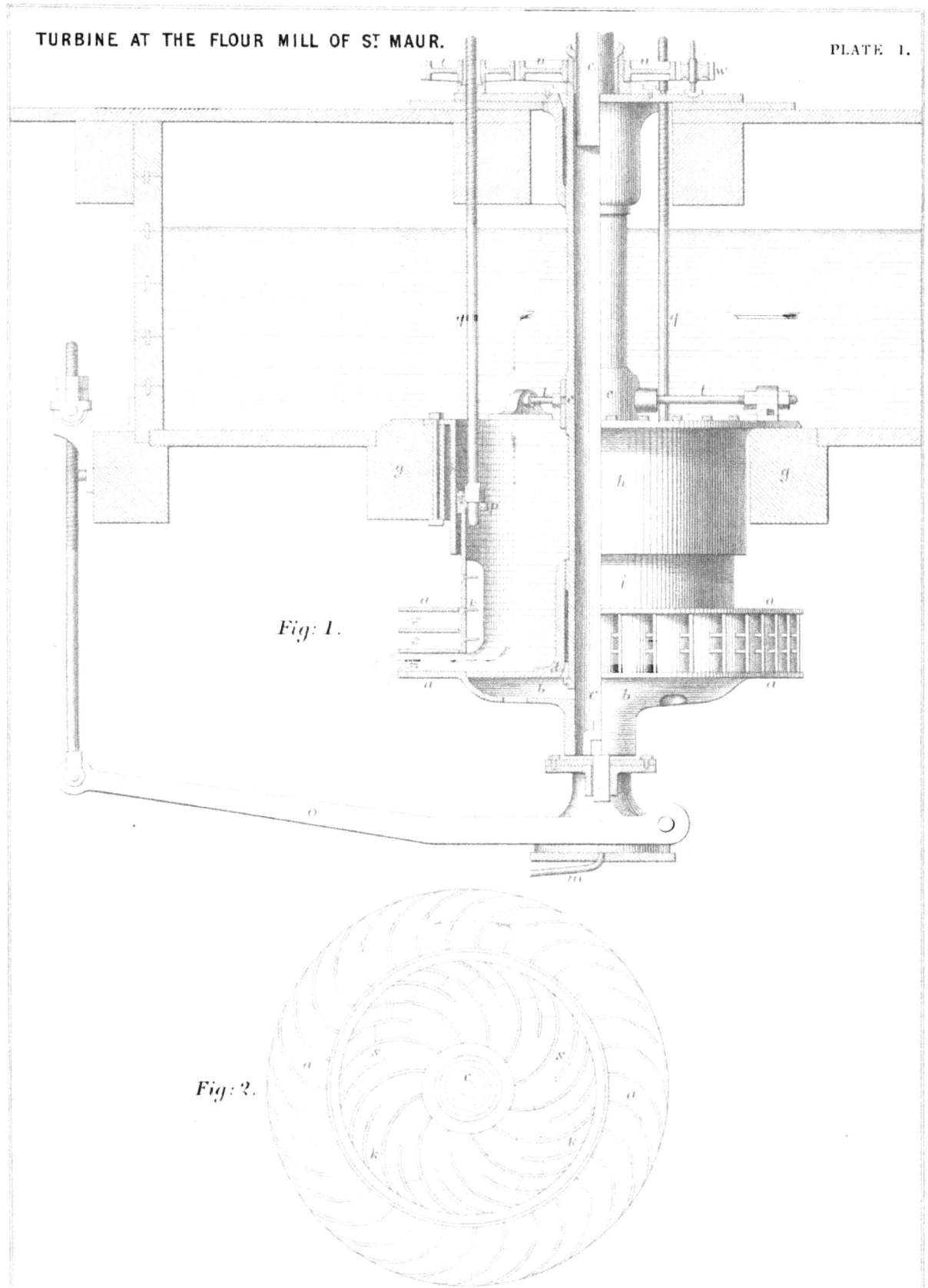

TURBINE AT THE FLOUR MILL OF ST MAUR. PLATE 1.

Fig: 1.

Fig: 2.

TURBINE AT THE SPINNING MILL OF ST BLAZIEN. PLATE II.

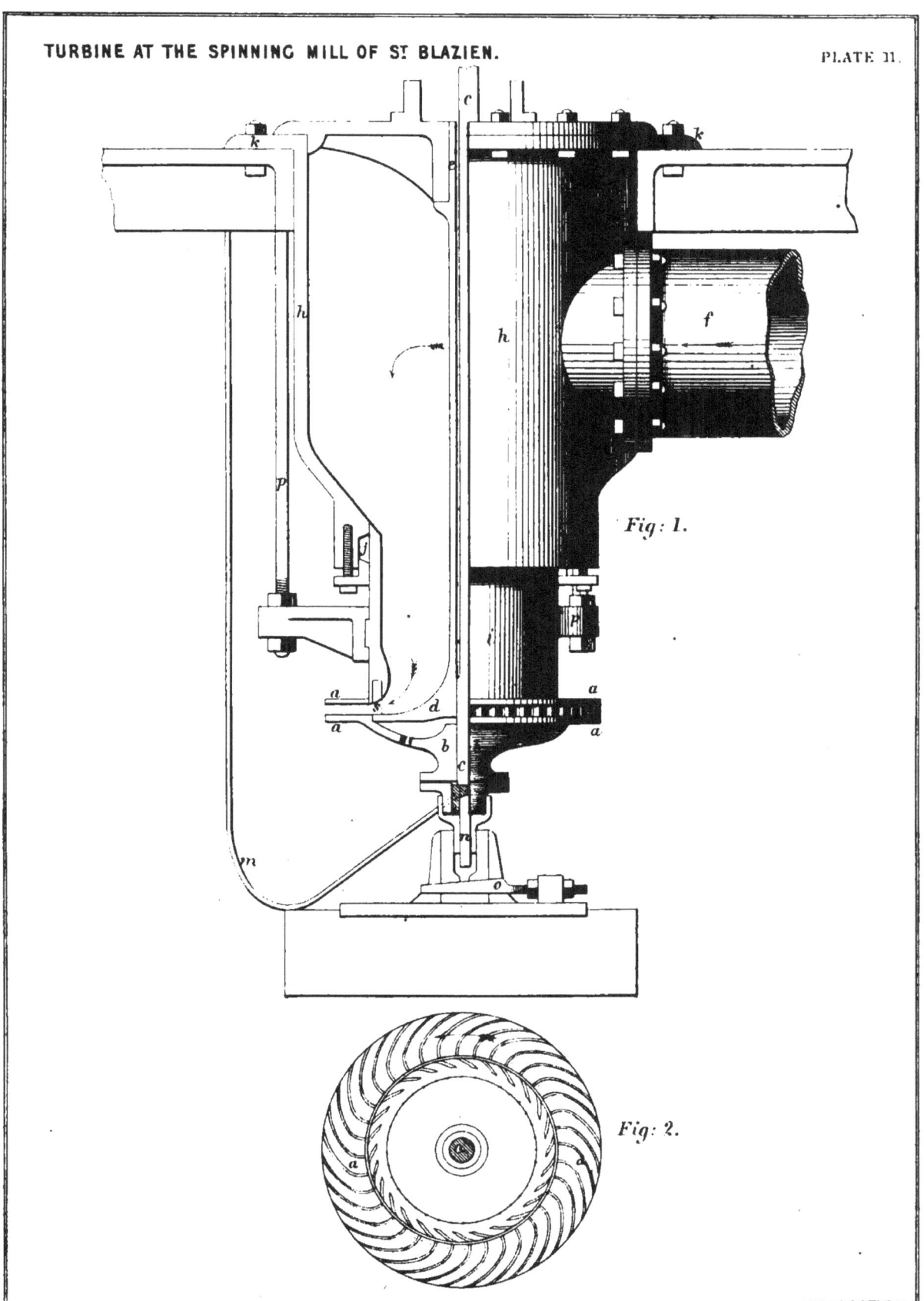

Fig: 1.

Fig: 2.

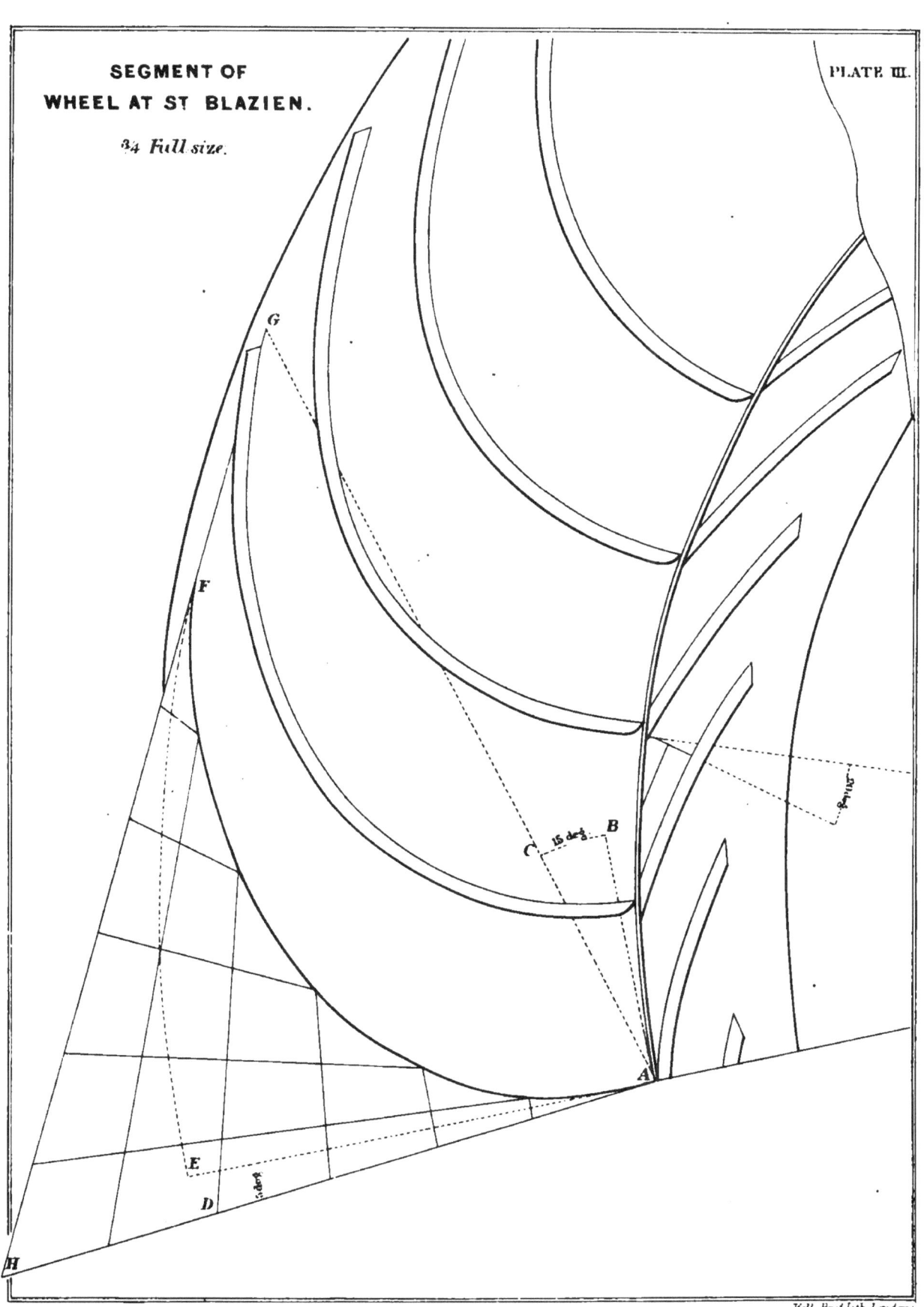

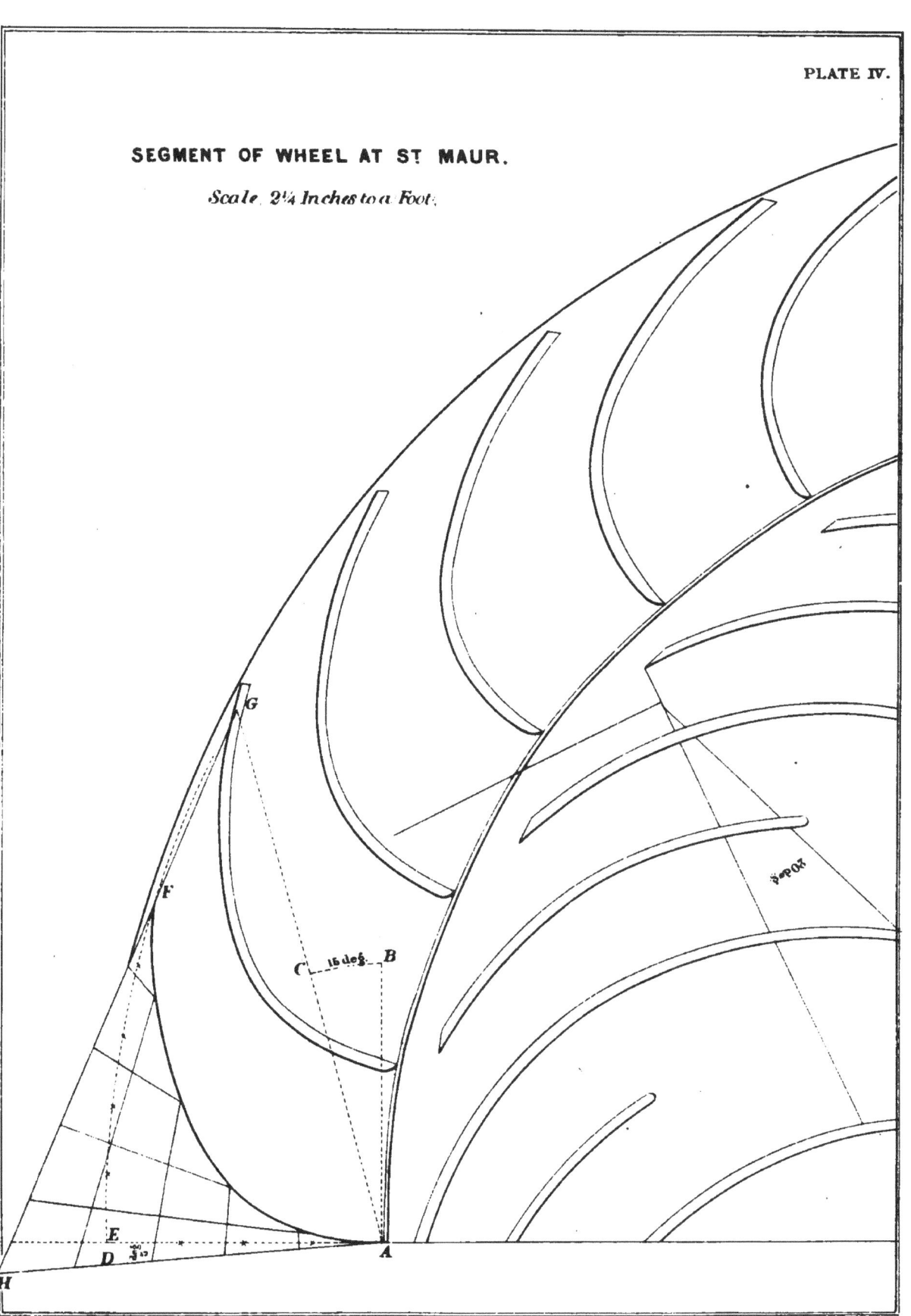

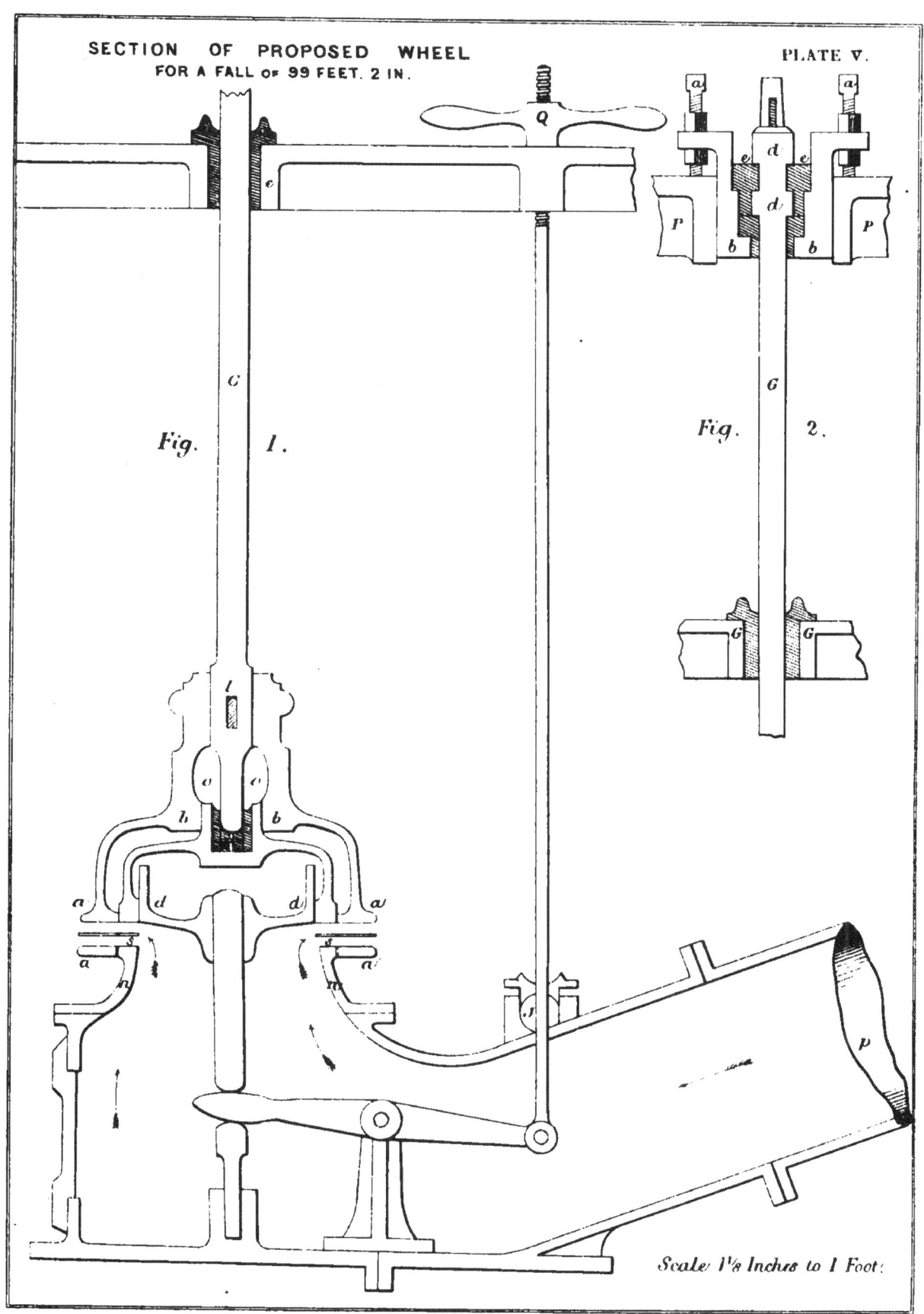

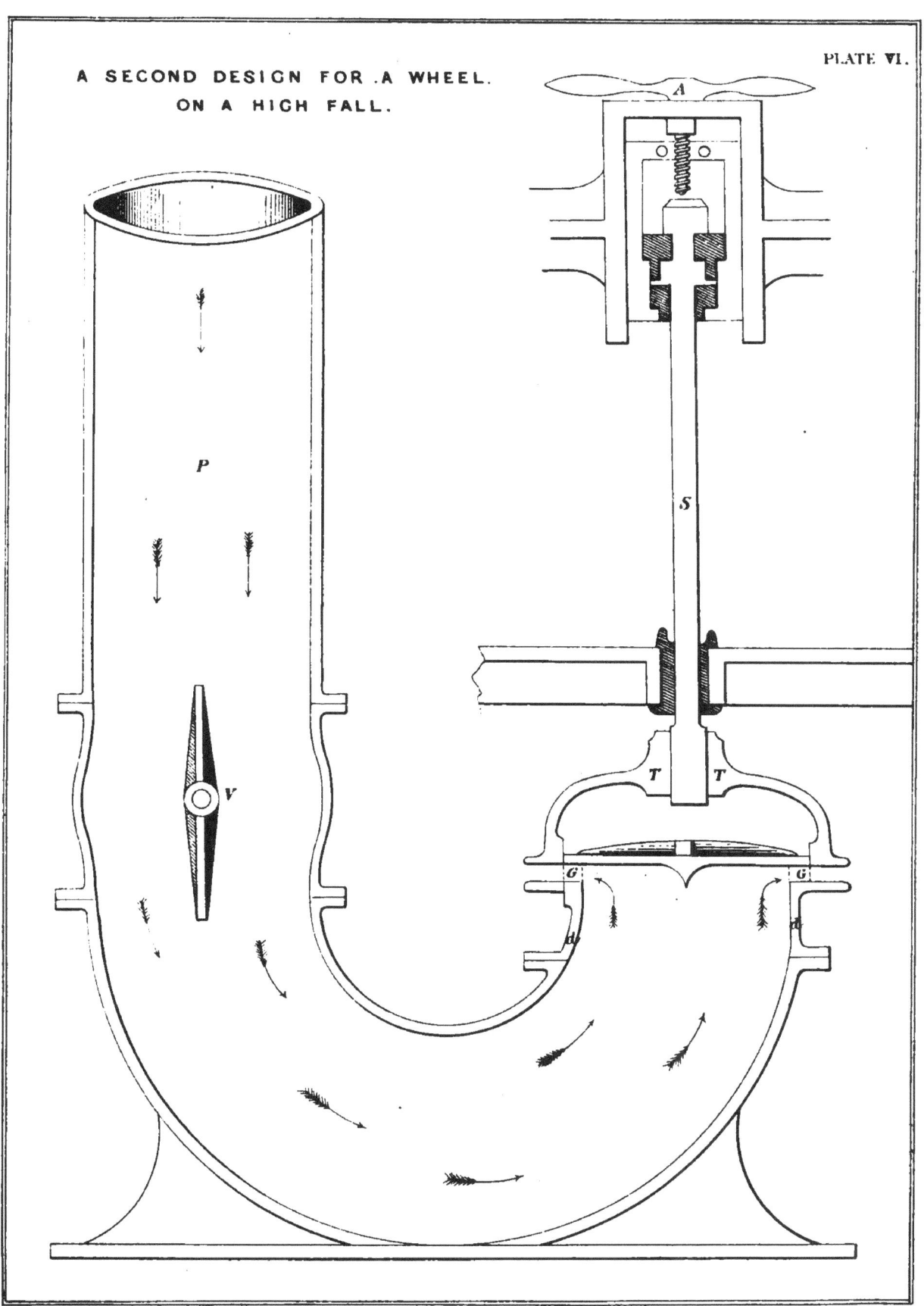

A SECOND DESIGN FOR A WHEEL ON A HIGH FALL.

PLATE VI.

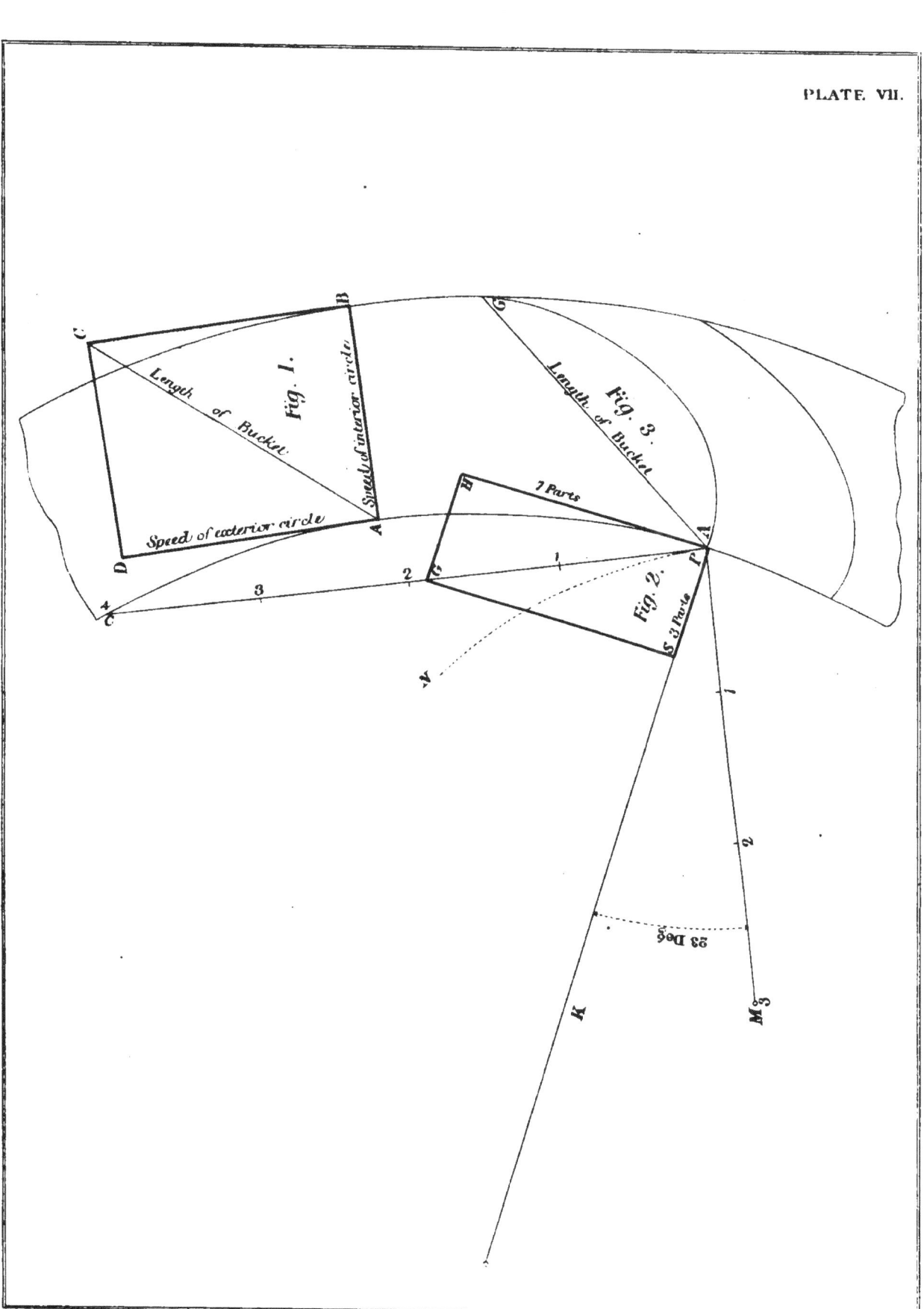

PLATE VII.

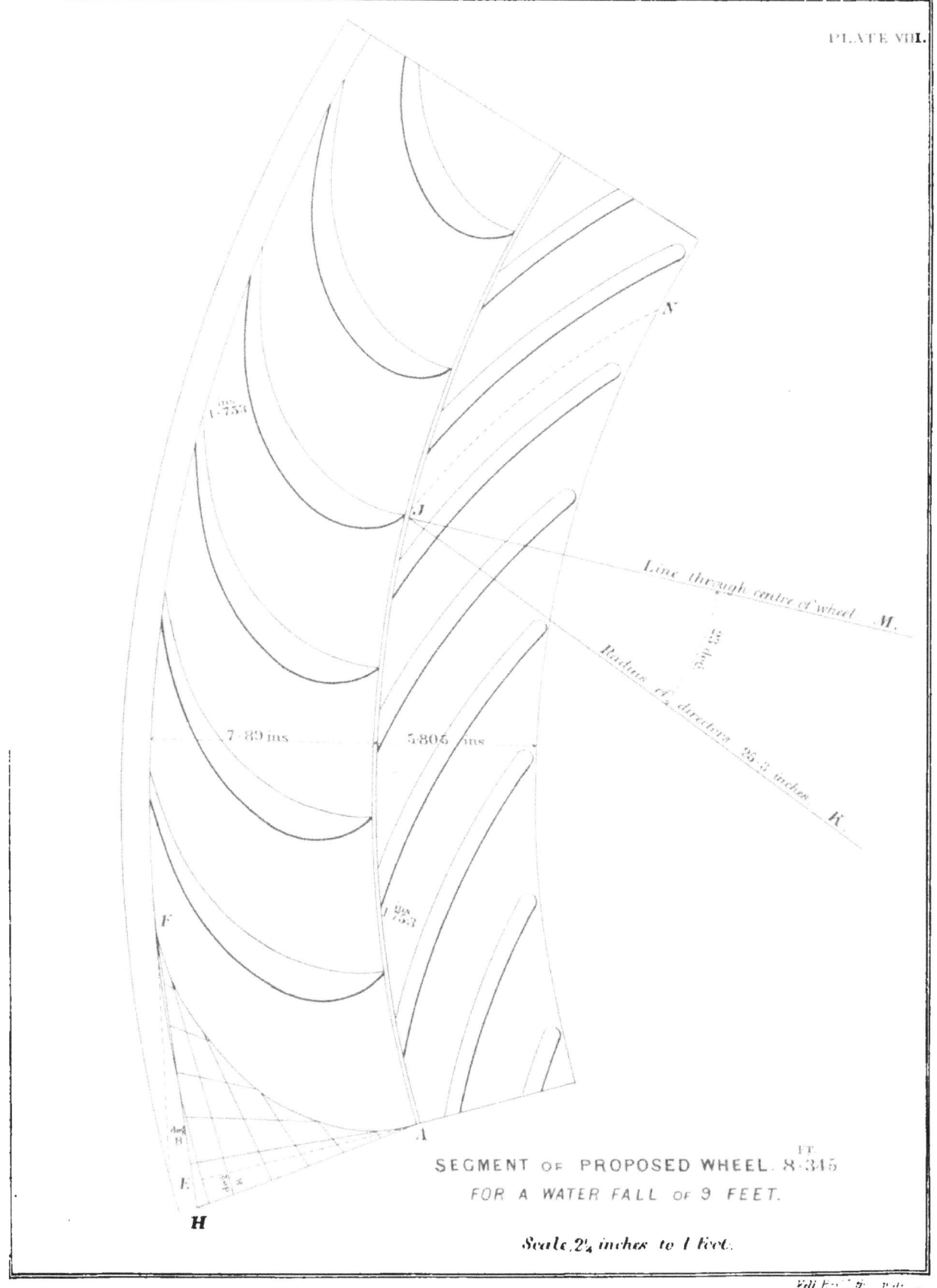

SEGMENT OF PROPOSED WHEEL 8·345 FT
FOR A WATER FALL OF 9 FEET.

Scale, 2¼ inches to 1 Foot.

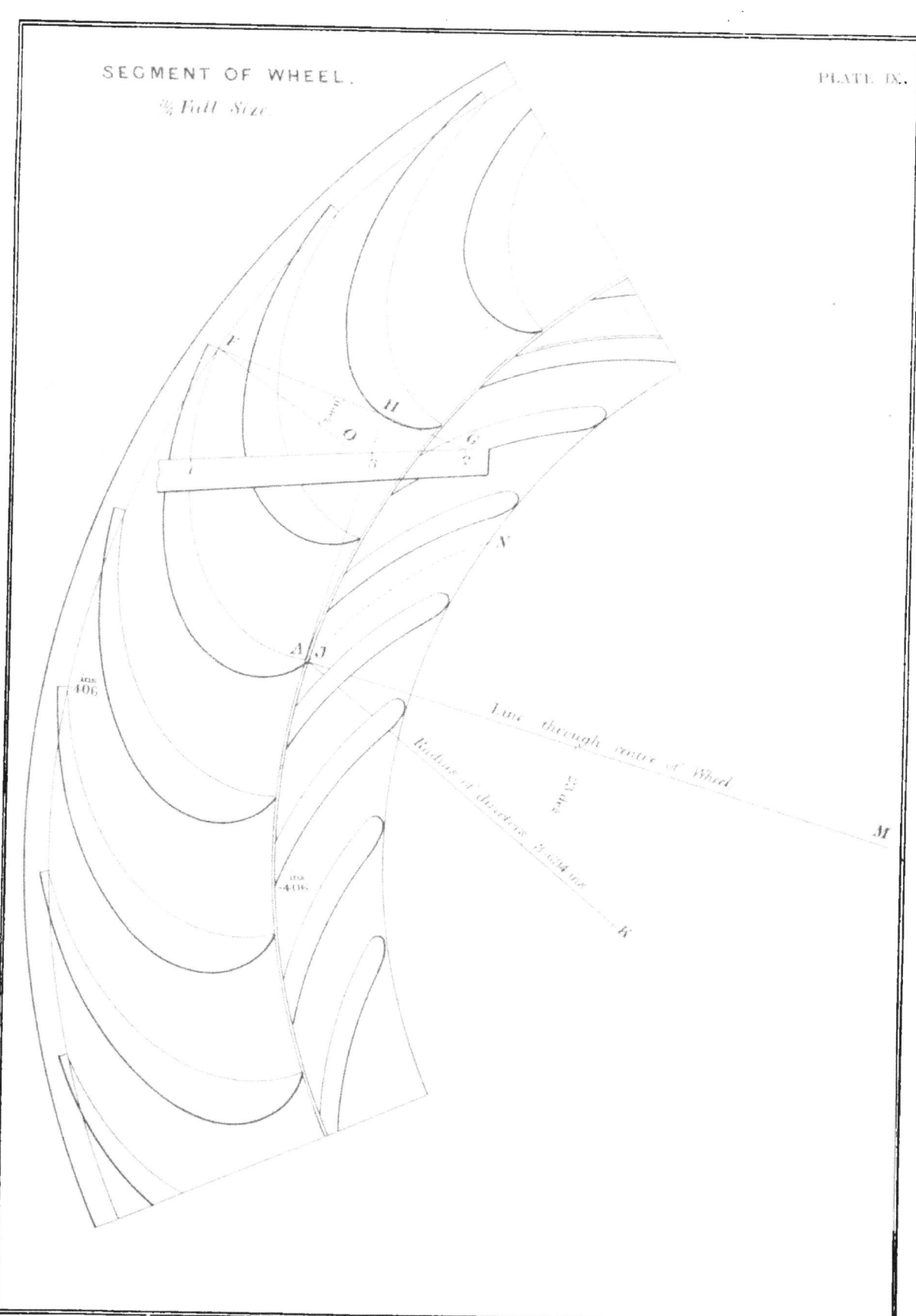

SEGMENT OF WHEEL.
⅜ Full Size.

PLATE IX.

PLATE X.

Fig. 1.   Fig. 2.   Fig. 3.

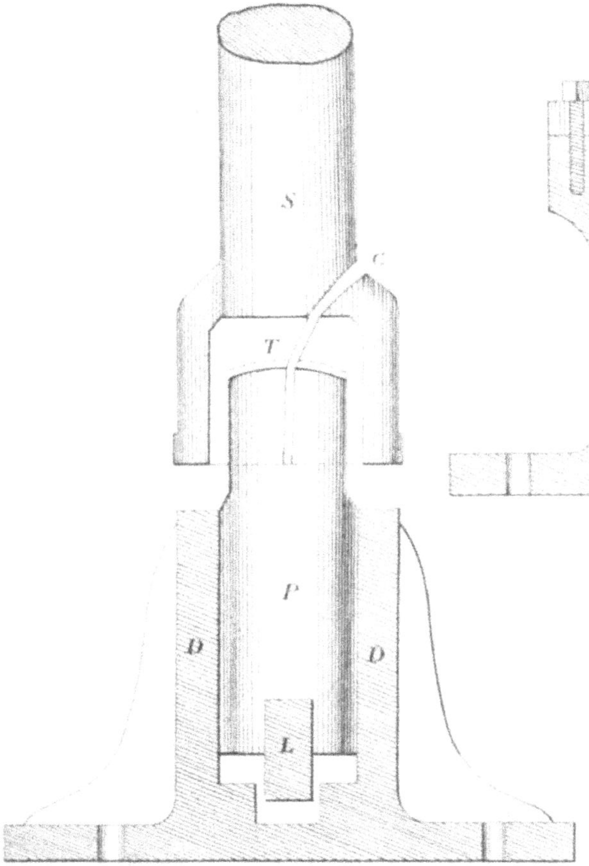

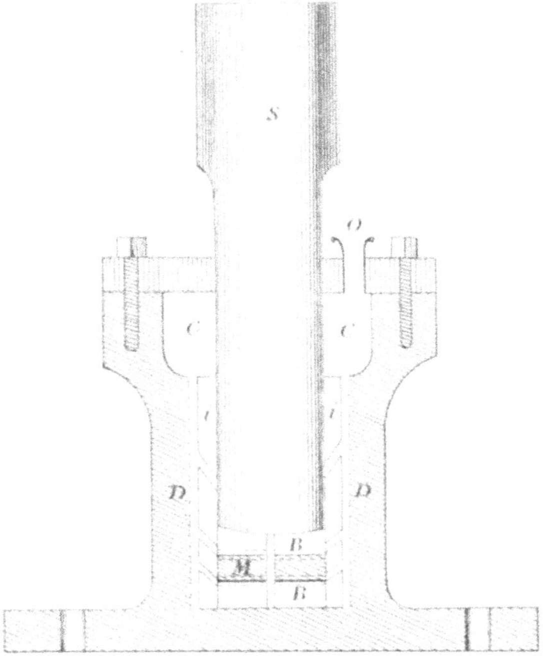

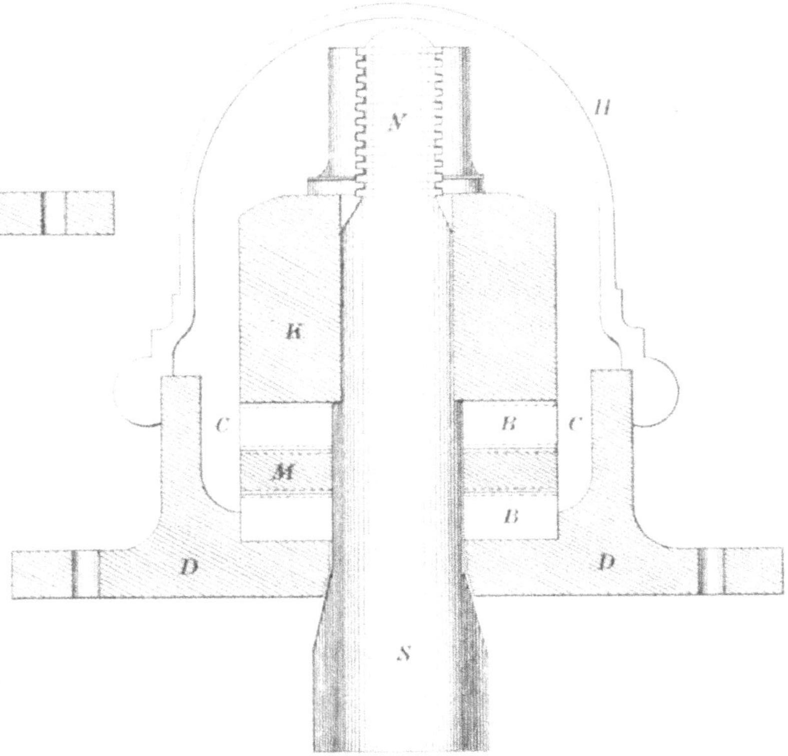

Scale 1½ Inches to 1 Feet

Kell Bros lith London

PLATE XI.

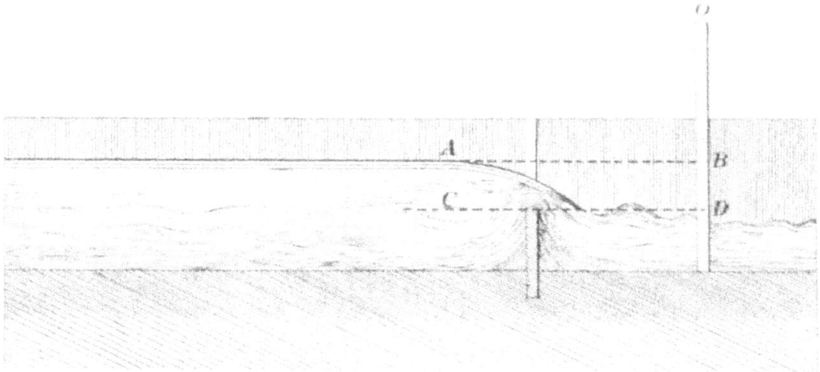

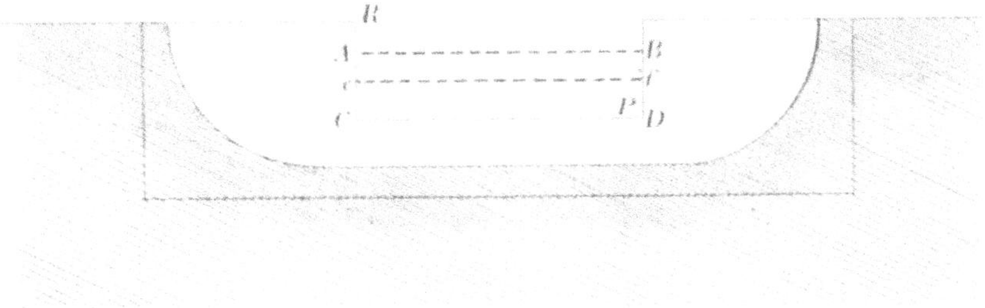

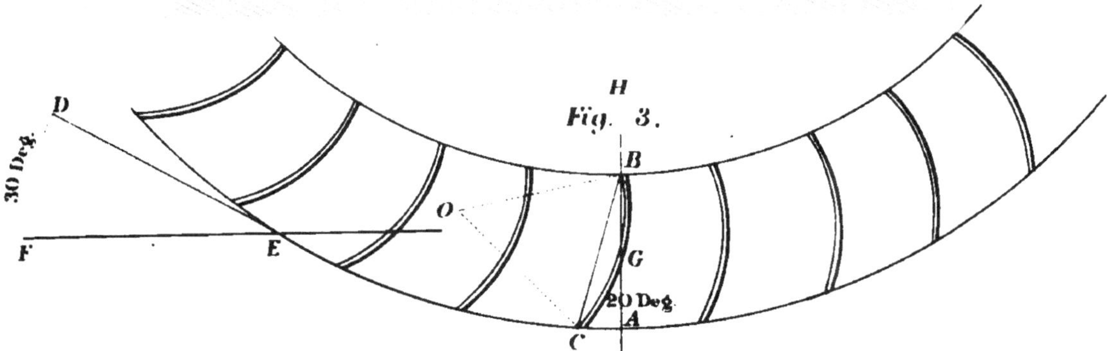

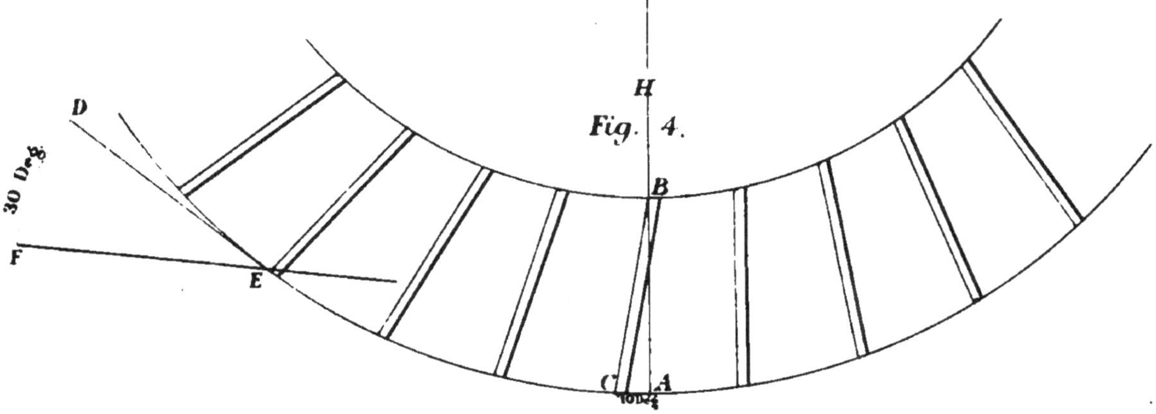

Fig. 3.

30 Deg.

20 Deg.

Fig. 4.

30 Deg.

10 Deg.

Scale for Figs. 3 & 4.
⅜ Inch = 1 Foot.

PLATE XII.

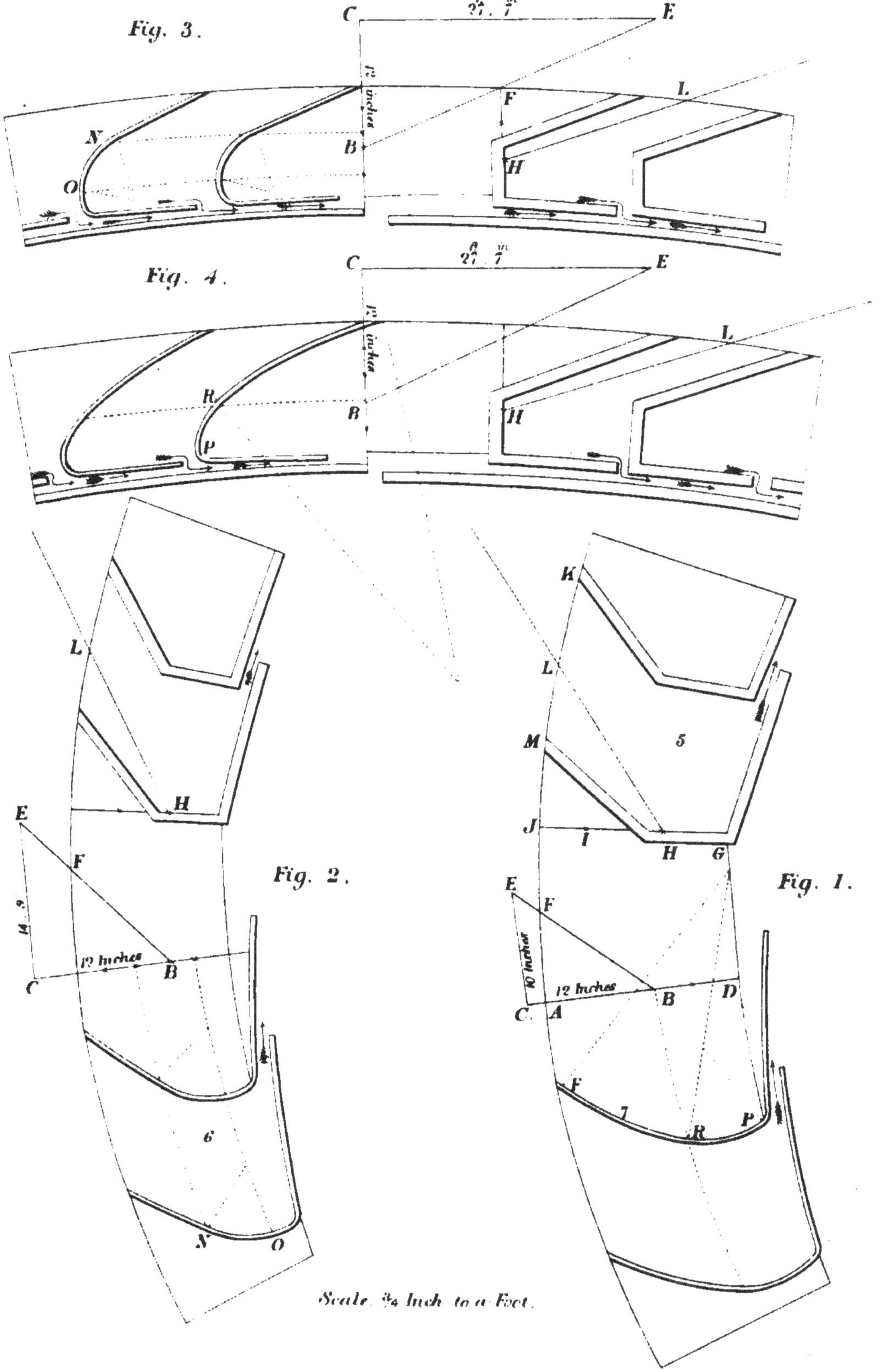

Fig. 3.

Fig. 4.

Fig. 2.

Fig. 1.

Scale ¾ Inch to a Foot.

www.ingramcontent.com/pod-product-compliance
Lightning Source LLC
Chambersburg PA
CBHW082352220526
45470CB00008B/2722